SECRET ANCHORAGES OF BRITTANY

Peter Cumberlidge

Published by
Imray Laurie Norie & Wilson Ltd
Wych House The Broadway
St Ives Cambridgeshire PE27 5BT England
℡ +44(0)1480 462114
Fax +44(0)1480 496109
www.imray.com
2016

All rights reserved. No part of this publication may be reproduced, transmitted or used in any form by any means – graphic, electronic or mechanical, including photocopying, recording, taping or information storage and retrieval systems or otherwise – without the prior permission of the Publishers.

© Peter Cumberlidge 2016
Aerial photography © Patrick Roach and Imray Laurie Norie & Wilson Ltd 2016

Peter Cumberlidge has asserted his right under the Copyright, Designs and Patents Act 1988 to be identified as the author of this work.

1st edition 1993 (Waterline Books)
2nd edition 2005
3rd edition 2016
A catalogue record for this book is available from the British Library.

ISBN 978 184623 815 4

This product has been derived in part from material obtained from the UK Hydrographic Office with the permission of the UK Hydrographic Office, Her Majesty's Stationery Office, Licence No. GB AA - 005 - Imray
© British Crown Copyright, Secret Anchorages 2016
All rights reserved.

CAUTION
Every effort has been made to ensure the accuracy of this book. It contains selected information and thus is not definitive and does not include all known information on the subject in hand; this is particularly relevant to the plans, which should not be used for navigation. The author believes that his selection is a useful aid to prudent navigation, but the safety of a vessel depends ultimately on the judgement of the navigator, who should assess all information, published or unpublished.

THIS BOOK IS NOT TO BE USED FOR NAVIGATION.
NOTICE: The UK Hydrographic Office (UKHO) and its licensors make no warranties or representations, express or implied, with respect to this product. The UKHO and its licensors have not verified the information within this product or quality assured it.

PLANS
The plans in this guide are not to be used for navigation. They are designed to support the text and should at all times be used with navigational charts. All bearings are degrees true.

Reprinted in Croatia by Denona, 2021

CONTENTS

THE PLEASURES OF ANCHORING, 6

ANCHORING IN BRITTANY, 15

Chapter 1
CAP DE LA HAGUE TO
ANSE DE PAIMPOL, 22

Chapter 2
ANSE DE PAIMPOL TO
THE PENZÉ RIVER, 54

Chapter 3
ILE DE BATZ TO L'ABER-ILDUT, 96

Chapter 4
CHENAL DU FOUR TO
RAZ DE SEIN, 122

Chapter 5
RAZ DE SEIN TO BAIE
DE LA FORET, 160

Chapter 6
AVEN RIVER TO PRESQU'ILE DE
QUIBERON, 198

Chapter 7
BELLE ILE TO THE GULF OF
MORBIHAN, 228

Chapter 8
HOUAT AND HOEDIC TO THE
VILAINE RIVER, 256

Chapter 9
THE VILAINE TO THE LOIRE, 270

INDEX, 291

PREFACE TO THE THIRD EDITION

In this third edition of Secret Anchorages, I have included many new pictures taken during our recent cruises around Brittany's fabulous coastline. I have also added some new anchorages and incorporated various suggestions from readers who have been kind enough to write to me with their own experiences of exploring these fascinating places.

As with previous editions, the book focuses entirely on natural anchorages and only mentions marinas or *ports de plaisance* in passing. To my mind, one of the true luxuries of cruising with your own boat is that exquisite feeling of independence and privacy you enjoy by spending lazy hours and days swinging to your own anchor, beholden to nobody and, in Brittany at least, rarely incurring dues. This wonderful, liberating sensation surely lies right at the heart of boat ownership, whether sail or power. It is really the floating equivalent of having a comfortable detached house with private watery views.

So for me, one of the most important characteristics of a rewarding cruising area is a tempting choice of anchorages, either beachy daytime stops for lunching and swimming, or more sheltered hideaways where you can spend a tranquil night or two in beautiful surroundings whatever the weather. More than anywhere else in Europe, Brittany has these advantages in spades, with its fantastic range of grand estuaries, rocky inlets, sleepy rivers, wild secluded bays and tiny, sometimes uninhabited islands.

On the north coast especially, the scene is enhanced by the extravagant tides which change the shape and atmosphere of the coastline almost as you watch. At high waters you can nudge into 'green-tinted' backwaters with plenty of depth under your keel, or meander up winding rivers to find peaceful creeks. On coastal hops you can safely cut across charted shoals or sneak inside reefs and islets. Low waters are even more enchanting, when many anchorages are perfectly cocooned by exposed rocks. Then you can lie in idyllic lagoons, run the dinghy ashore onto clean sand and dabble among rock pools in nostalgic shrimping net bliss.

The aim of this book, therefore, is to provide concise pilotage directions for as many natural anchorages as possible along the whole Brittany coast, from the Bay of Mont St Michel up in the far north-east right round to the border with La Vendée. The coverage actually starts near Cap de la Hague on the Cotentin Peninsula and includes Iles Chausey, both of course part of Normandy. Then we enter Breton waters at Cancale and follow the coastline westwards to Ushant, south via Brest and Douarnenez into Biscay, and all the way round – for over 300 miles – to the majestic mouth of the mighty Loire, anchoring in every possible nook and cranny on the way.

The more remote coastal anchorages hardly change at all over the years, or only very slowly. While certain prominent features ashore may become obscured by trees or occasionally by new buildings, the cliffs, headlands, navigation marks and elegant lighthouses around Brittany remain remarkably constant. Some anchorages covered in the first two editions of this book have gradually filled up with local moorings, leaving less room for swinging to your own hook or forcing you to anchor further out where the shelter isn't so good. Other spots have actually become less crowded with buoys since the first edition was written, as changing berthing tastes and marina expansion have tempted boats away from swinging moorings and slotted them into neat rows of pontoon fingers.

Anyone with even a pinch of salt in their veins will never tire of Brittany, even when cruising between gregarious harbours and marinas. But by anchoring in natural havens that have been used from time immemorial by those whose livelihood turns on the rhythm of the tides, you can tune more finely into the spirit of this special part of France.

The seafood here is another prime attraction. There are oyster and mussel beds all around this coast, and sturdy Breton fishing boats are continuously working offshore to bring home the crabs, lobsters and langoustines expected on menus every day. On the north coast, especially in the Bay of St Brieuc, they harvest succulent scallops – Coquilles St Jacques – that are second to none. And at low spring tides, with a few simple tools, you can join locals scavenging among the drying flats and rocky ledges for mussels and clams.

All these timeless delights are easily accessible from our own south and west coasts, and every summer the lure of Brittany is irresistible for the owners and crews of all kinds and sizes of boat. First-time visitors are usually astonished to find the spectacular seascapes completely unlike the English side of the Channel, while regulars are always grateful that these stunning cruising grounds never seem to change.

Peter Cumberlidge
September 2016

Stormalong

The rocky approaches to L'Aber-Ildut

THE PLEASURES OF ANCHORING

There is nothing to equal that magical feeling of peace and well-being that follows your safe arrival at a sheltered and secluded anchorage. The tensions of pilotage are over, your ship seems pleased to be at rest and you are poised, momentarily, between the sea and the land. It's good just to sit in the cockpit a while, adjusting gradually to your new surroundings and soothed, perhaps, by the sounds of the tide trickling past the hull, a slight swell collapsing on a nearby beach, or wind in the trees at the top of a cliff.

Generations of yachtsmen have savoured this timeless experience, whether at the end of a long ocean voyage, a Channel crossing, or a short passage round the coast from their home port. Eric Hiscock, for example, was cruising on the Brittany coast in 1939, only three weeks before the outbreak of war. This passage from his *Wandering Under Sail* always strikes a chord with me, especially when you consider that the approaches to the Tréguier estuary have probably changed very little in the years since these words were written:

'Then at last the sun came out, and as I took off my oilskin, lit my pipe and guided *Wanderer* up that lovely river in the rain-washed evening light, I tasted once again the greatest joy which small boat cruising can offer: the satisfying contentment of a rough and anxious passage successfully achieved. Quietly we wound our way between the woods and fields, and finally brought up in the little bay below the town of Tréguier. Only the deep tolling of the cathedral bell broke the silence as I put the last tier round the mainsail . . .'

Of course, marinas had not been invented then. Anchoring was a way of life and anyone coasting would always be keeping an eye on the large-scale chart for likely coves, river mouths or other protected spots for when the tide turned, or in case the wind shifted or looked like dying. In that less hectic era, and in common with most yachts of her type and size, *Wanderer II* had no engine, so her anchor would have been ready to let go whenever she was close inshore.

The pace and style of cruising is rather different now and our passage destinations are much more predictable and yet there is still tremendous satisfaction in nosing into a remote anchorage and feeling that you've arrived somewhere real. Marinas are now ubiquitous and all rather similar, like international hotels. One line of pontoons looks much like another, and if you shut your eyes and open them again in a marina, you could be almost anywhere. Yet despite this bland, unattractive sameness, it sometimes seems as if anchoring, at least proper overnight anchoring, is becoming a dying art.

In fine summer weather, to be sure, popular beaches and sheltered bays are often well filled with boats at anchor just for the day, since there's no better way to enjoy the holiday atmosphere of a sunny beach than from the cockpit of your own boat. But come the late afternoons, the mass exodus to nearby harbours and marinas usually starts in earnest. By early evening, only a few scattered boats may be left at anchor out of what, at lunchtime, may have been a sizeable fleet. As dusk creeps in the numbers will probably thin out further until only one or two shadowy hulls are still swinging

Peacefully dried out in Port de la Corderie

placidly to their own hooks, their crews savouring that luxurious sense of freedom and independence that comes from lying overnight at anchor aboard your own boat.

This experience can be just as relaxing in brisk or stormy conditions as during calm summer spells, so long as you are nicely tucked under a weather shore with the wind howling safely overhead. To be anchored snugly in a sheltered pool under the lee of the land while it's blowing a hooley outside is one of the most restful aspects of cruising I know. Just pass me a good book and a whisky and soda to make the contentment complete. George Millar's *Oyster River*, a classic cruising tale about a long summer in the Gulf of Morbihan, has many good descriptions of idyllic anchorages, including this brief memory of a stormy night spent at anchor near the head of the Morbihan:

'We gybed round in the middle of the narrows and anchored opposite several hulks of tunnymen rotting on the beach. I put off at once in the dinghy, mooring her to an anchor upstream as well as down so that she could not wander when the tide turned.

'At sea it was a night of terror with a severe gale from the west and torrents of rain. Yet we lay there in peace, only aware of the wind by its howling in the trees of the Aradon bank just above us.'

In this extract you can appreciate the sense of freedom about anchoring that existed in the early 1960s, when there were few moorings in the Morbihan. Many of the most attractive spots are crammed with moorings now, and yet there are still secret corners to be found, especially in the shallow eastern reaches or up one or two unlikely-looking inlets that seem on the chart to have more mud than water. The Anse de

Anchored in the Trieux estuary

Badène is a good example, a little visited tongue of shallow water opening off the east side of the Auray River.

But quite apart from the fact that all the most attractive cruising areas have become more crowded over the years, you can also see a trend towards more trammelled cruising that is to some extent self-imposed. Because each season, little by little, our traditionally liberating pastime becomes just that bit more regulated, that bit more gregariously organised as a leisure market rather than an easy-going means of real escape from routine, it can be all too easy to slip into a kind of 'Butlins' mentality, following the general cruising drift from one set of visitor berths to the next. Taking this habit to extremes, you may as well be on a watery package tour, following the set itinerary as you look for the allocated pontoons, check in at the marina office, pay your dues and are issued with the current code number for loos and showers. Marinas have their practical advantages, but in high season they can seem little different in character from close-packed housing estates, with even less chance of escaping from neighbours.

I'm not averse to people, in fact some of my best friends are people, but something I definitely look forward to when cruising is a special quality of peace and seclusion that's becoming increasingly difficult to find on dry land, especially in overcrowded England. So squeezing into a crowded marina or rafting three or four abreast like a gaggle of Hong Kong junks never seems to me a particularly salubrious end to a sailing day. On the other hand, to fetch up in some secret cove or inlet to enjoy that first evening drink with room to breathe is one of life's true luxuries. And in France this five-star sensation is usually free of charge, which is an interesting reflection on economic forces. This being the case, even a couple of nights on the hook along the Brittany coast will pay for a splendid *menu gastronomique* when you do touch base again. It's always pleasant to think that your oysters and *vin blanc* are being financed out of funds that would otherwise have disappeared in marina dues.

Why quite so many cruising boats automatically head for marinas every night instead of sampling some of the comfortable freedom of natural anchorages is difficult to say, but the irony of this tendency is that most modern boats are actually extremely well adapted to convenient anchoring. The mechanics of the whole business are no longer so labour intensive, since many foredeck crews now only have to operate a switch to haul up their anchor and chain. Cruising boats of any size have comfortable showers and bathrooms, so you don't end up looking like Robinson Crusoe after a few nights away from civilisation. Efficient fridges mean that you can eat well, keep wine cool and not have to resort to long-life milk in your tea. Comfortable dinghies slung in davits make going ashore completely painless and easy-to-use stern boarding ladders allow you to hop in and out of your tender without fuss. All in all, we've never had it so good.

Of course it's attractive to be able to wander ashore straight off your boat into fine old towns such as St Malo, Tréguier, Concarneau or Vannes. But it's also fascinating to land with the dinghy on an empty beach or at some forgotten stone jetty in the peaceful reaches of a Brittany river. Exploring ashore by the back door, you can more easily start to absorb something of the real character of Brittany away from the well-trodden tourist trail.

Some skippers seem to feel more secure when attached to a pontoon for the night and perhaps sleep more soundly in their bunks for not having to worry about the possibility of a dragging anchor or a sneaky wind shift in the small hours. Many newcomers to cruising have simply learnt to regard marinas as right and proper places to make for at the end of the day, convenient bases where they can top up with water or diesel, plug into shorepower and keep the fridge running so the drinks stay cold. Anyone with a young family or restless teenagers aboard will be attracted to marinas for obvious reasons, although youngsters in their less demanding pre-teen years can have plenty of fun with the dinghy while lying at anchor, picking up a great deal in the process about boat-handling and basic seamanship. I remember nothing quite so idyllic in my own childhood as being off and away in the dinghy, landing at what seemed to be uninhabited islands or private beaches, and exploring rivers, shallow creeks or rocky inlets from the base camp of some anchorage that felt like the Amazon or Orinoco rather than Brittany.

On the question of security, it's worth remembering that, in certain conditions, negotiating the close confines of a marina can be much more hazardous than lying at anchor. There are marinas in Brittany rivers where the tide runs strongly through the pontoons, making it tricky and potentially dangerous to manoeuvre in and out. Tréguier is a good example on the North Brittany coast, where the river current pours at a slight but devilish angle past the pontoon fingers at well over three knots on a spring ebb. In such circumstances it's only safe to enter or leave the marina at dead slack water. Over the years I've seen some gruesome incidents at Tréguier where boats have tried to manoeuvre while the tide was flowing strongly.

I recall one particularly dramatic collision when a 33ft yacht was swept rapidly sideways after reversing boldly out of her berth but then dithering a little about turning and getting under way. I watched her carried inevitably and relentlessly down onto the end of a downstream pontoon finger and had a hole neatly punched in her side, luckily above the waterline. The outer marina at Lézardrieux can also be a trap for the unwary with a spring tide running, although the streams here are not as powerful as at Tréguier. The pontoons at Roscoff and Bénodet suffer from similar effects. Not only do you risk damaging your own boat when manoeuvring in such conditions, but you might also incur substantial third party claims from anyone you accidentally hit. Compared with such tensions, the pleasant simplicity of lowering or raising your anchor with plenty of room all around you seems highly attractive and considerably less dangerous.

Anchoring can provide strategic advantages when you are working favourable tides around an indented coastline. If you imagine cruising westwards along the coast from St Malo, which is right at the east end of North Brittany, it can be both pleasant and efficient to reach L'Aberwrac'h – at the far west end – in three easy stages, anchoring overnight at the end of each stage. Timing the first leg to carry a full period of ebb from St Malo, you might run out of fair stream somewhere near Ile de Bréhat and be therefore well placed to fetch up in Port de Guerzido or La Corderie. From either of these anchorages you could easily set off westwards again at high water next day to round Les Héaux and carry another full six hours of fair ebb to arrive near low water off the Morlaix estuary. A choice of several useful anchorages awaits hereabouts, allowing you to enjoy the rest of the day in peaceful surroundings, conveniently positioned to press on again with next day's west-going ebb.

Anchoring in the lower estuary or the Chenal de Batz would be much simpler than having to follow the Morlaix River six miles inland near high water, lock into the

Port de la Corderie, Ile de Bréhat

marina basin there and then have to lock out again next morning before dropping back downriver. What could be easier, on the other hand, than to leave an estuary anchorage at high water next day and set off straightaway towards L'Aberwrac'h, using a full ration of fair tide and aiming to arrive off the Libenter entrance buoy nicely before the next foul flood sets in.

There are many locations around the Brittany coast where, in the right conditions, you can save time, distance and harbour dues by anchoring just inside a river mouth rather than motoring well inland, sometimes for several miles, to a marina. If, as skipper, you are keen to press on next day, perhaps if time is running short after a cruise, not only will that saved distance be an advantage in itself, but you'll also find it easier to mobilise your crew next morning and get away somewhere near your planned time. While you are lying at anchor, there's no risk of 'losing' any of your crew in last minute shopping, visits to cafés or lingering sojourns in the marina showers. Naval captains in the days of sail understood this principle only too well and always favoured anchoring in a sheltered roadstead to being warped alongside a quay, where their crews could more easily vanish into the fleshpots ashore.

Apart from the practical pleasures and advantages of anchoring overnight when cruising, there is also a definite aesthetic appeal. This complex sensation has its roots partly in a taste for self-sufficiency and sometimes in a vague but tangible appreciation of maritime history and seafaring tradition. There's something rather satisfying about following in the navigational footsteps of past mariners who, working completely under sail with no recourse to engine power, used natural anchorages as a matter of course when working their passages. The skippers of Brittany sailing coasters, fishing boats or privateering luggers would know every

Anchored off Plage de Trescadec, Audierne

nook and cranny along their local coasts where it was possible to drop the hook in different combinations of wind and tide. They would always have a plan about where to fetch up if they ran out of fair stream, ran out of breeze, needed shelter from a rising wind or had to find somewhere safe to wait out a fog. They'd know how safe or easy it would be to leave particular anchorages at night, in poor visibility or at different states of tide. They would instinctively realise which spots were best at springs and which most suitable at neaps.

While all this intricate understanding is less vital in practice today, it nevertheless remains a fascinating part of fully appreciating a cruising coastline. You can soon learn a great deal about the geography and character of any stretch of coast by developing the habit of anchoring overnight. And if, in some deserted bay or inlet, you row ashore, climb to some vantage point and look down on your own boat swinging to her own chain, you'll feel stirrings of those same sensations felt by sea captains and explorers through the ages. It might be Ile de Bréhat, the Morlaix River or Anse de Blancs Sablons. Just for a moment though, it could almost be Darien or Botany Bay.

My intention with this new edition of *Secret Anchorages* has been to include directions for as many natural anchorages as possible around the whole Brittany coast, starting from near Cap de la Hague (which is actually outside Brittany) and working west and south round the Brest peninsula, into the Bay of Biscay and down as far as the Loire. Even with the extra anchorages I have added to this edition, I'm sure I won't have picked up every feasible hideaway on this intricate coast, so any useful additions or corrections will always be gratefully received, if you could spare the time to drop a short note to Imray Laurie Norie and Wilson Ltd at Wych House, The Broadway, St Ives, Cambridgeshire PE27 5BT. *Email* ilnw@imray.com

I should also point out that this book is not a pilot in the usual sense, in that it covers only those natural havens where you can lie to your own anchor. *Secret Anchorages* aims to visit those spots that the pilot book, for simple reasons of space, cannot reach; it also extends well down the Biscay coast, thereby covering North, West and South Brittany in the one volume.

Some of the anchorages will only be safe under certain conditions of wind and weather; some will be tenable at neaps but not at springs; some are restricted to shoal-draught boats, or to keel-boats that can rig drying-legs. While I have tried to give guidance on these factors, I have also aimed for brevity in the directions, in order to be able to include as many anchorages as possible. I have assumed throughout that readers will be cruising with a good selection of large-scale Admiralty charts, and I have listed those charts I consider most useful for each section of coast covered by the book. In some cases, the French equivalent of our Admiralty Charts, published by the redoubtable Service Hydrographique et Océanographique de la Marine (SHOM), are preferable for their large-scale coverage of a particular area, and I have recommended these SHOM charts where appropriate.

When anchoring off the North Brittany coast in particular, the considerable range of tide needs to be taken into account, both for the obvious reason of allowing sufficient scope of cable, but also because the shelter an anchorage affords from swell may be markedly different at different states of tide. In many North Brittany anchorages, you'll get the best shelter near low water, especially at springs. Then, you can often find yourself almost land-locked by a cordon of natural breakwaters – drying rocks grandly exposed on the ebb to provide increased protection from swell. By the same token, if you arrive at an anchorage near low water and all seems snug, be prepared for an increase in motion as the flood comes back, especially for the two hours either side of high water.

One important question to ask about any natural anchorage is whether it's feasible to leave at night if a wind shift should force you to clear out. In some cases I have indicated where an anchorage is completely safe to leave at night or, conversely, where it is dangerous to do so.

When anchoring in rivers or shallow bays anywhere on the Brittany coast, it's important to steer well clear of oyster or mussel beds, of which there are many. Shellfish cultivation represents a significant part of the Brittany economy, and a hefty great CQR or a couple of iron bilge-keels settling on an area of young oysters will not endear you to the local population. Oyster and mussel beds are usually shown on large-scale charts and will probably be marked in any case by forests of withies. There are important areas of shellfish beds off Cancale; in the Baie de la Fresnaie just east of Cap Fréhel; in the Anse de Paimpol and parts of the Trieux estuary; in the Tréguier River; in the Morlaix and Penzé estuaries; at L'Aberwrac'h not far above the marina, and in the upper reaches of the river above Pointe Cameuleut; in the upper reaches of L'Aberbenoît. Take great care when deciding where to fetch up, and don't forget that those tempting menus of *fruits de mer* have all got to be supplied from somewhere.

Throughout this book I have been keen to emphasise that anchoring in a sheltered cove or inlet for the night can bring the obvious benefits of greater seclusion, peace and quiet, and freedom from men with receipt books asking for harbour dues. Yet it also offers something more intangible, a return to that sense of being in touch with the sea and the weather which is somehow lost when you secure to a pontoon, plug in the shorepower and step ashore to another marina complex with facilities laid on. Swinging to your own anchor and cable, you can somehow gauge more easily those subtle shifts in wind strength and direction that may bear upon the strategy for tomorrow's passage.

Sometimes, of course, one can be too much in tune with the elements, perhaps if a weather shore has become a lee shore overnight, your ship is rolling in protest and you are trying to snatch a bite of breakfast before being forced to get under way soon after dawn. Then you start to crave for a sheltered *port de plaisance* near a sleepy Breton town, with showers and hot croissants close to hand, and a friendly smile from the Capitaine as you settle your modest account for a carefree cruising night.

High water in peaceful Port de la Corderie on the west coast of Ile de Bréhat

ANCHORING IN BRITTANY

While the essential skills, equipment and judgment about anchoring apply to virtually any cruising area in the world, different coastlines have different quirks and characteristics that need to be appreciated if you are to get the best from exploring their anchorages. This book covers natural anchorages right around the long Brittany coast and indeed starts off with a short stretch of adjoining Normandy, so it's worth considering some of the special features that are more or less relevant to different parts of this area.

THOSE FRIENDLY ROCKS

The north and west Brittany coasts are renowned for their copious rocks that sometimes straggle offshore for several miles. The sheer extravagance of these reefs and shoals can be alarming to first-time visitors, especially when you consider the dramatic tidal range and powerful streams of North Brittany in particular. But after a couple of seasons cruising these spectacular waters, navigators soon come to regard the rocks of Brittany as helpful friends as well as potential hazards. For one thing, their shapes and positions never change, at least not noticeably in our fleeting cruising lifetimes. They can therefore serve as reliable pilotage marks, especially when you get to know their profiles and moods and gradually become less nervous about seeing such apparently menacing reefs at relatively close quarters. Once you are familiar with the idea of craggy rocks jutting up all over the place, your large-scale Brittany charts become doubly fascinating as you start to identify useful features in detail. Rocks can provide valuable impromptu transits that can help you enter coastal anchorages safely, guide you towards the best spots to drop your hook and, when you have fetched up and settled back on the chain, serve as vital cross-checks that your anchor is holding well.

Brittany's famous rocks also provide welcome shelter for many coastal anchorages, especially as the ebb falls away to expose more and more granite heads that gradually join up to form perfect natural breakwaters. This tidal factor is important for many Brittany anchorages, having a marked effect in increasing or reducing the shelter they offer. It's worth noting that spring low waters along the north coast occur in the early afternoons and the early hours of the morning. This means that anchorages that are naturally protected by cordons of reefs are generally at their quietest just after lunch at spring tides – which is ideal for holiday lazing – and again at the best time to contribute to a peaceful night's sleep.

SPRINGS AND NEAPS

Exploring Brittany anchorages is usually most fascinating around springs, especially along the east half of the north coast where the mammoth tidal range creates spectacular coastal seascapes when the ebb runs away. To be anchored in Chausey Sound, the Anse de Paimpol or around Ile de Bréhat at low springs is a fantastic

Anchored in Chausey Sound

experience spiced with all the elemental grandeur of this magnificent coast. It seems incredible that the sea will ever return, as you gaze across miles of exposed reefs in Chausey, acres of mussel beds in the Paimpol approaches or that veritable moonscape of granite that, at low tide, shrinks the channels around Bréhat to a quarter of their high tide width. A low spring tide can provide almost perfect protection if you are anchored right in the middle of such a scene in a virtually enclosed pool. Of course the corollary of this is that a spring high water submerges most of your granite breakwaters for several hours and lets any rolling scend come filtering in.

Neap tides also have advantages for anchorage hopping. In areas of large tidal range, low water at neaps is significantly higher than low water springs, which means that cruising at neaps opens up many more anchorages in which you can stay safely afloat. Neaps also allow you to anchor much further into certain coastal inlets to obtain better shelter. Port Blanc is a good case in point on the North Brittany coast, Ile Melon on the west coast and the Rivière de St Philibert on the south coast. At dead neaps, most boats of moderate draught can edge well into these three fascinating hideaways and still stay afloat at low tide, whereas at springs the relentlessly emptying ebb forces you to anchor further out in potentially more rolly water.

Of course neap low waters don't expose quite the same expanses of coastal rock as emerge at springs, so your shelter from swell is that much less effective. At neaps you also have to be a bit more careful around high water, because dangers that you might be used to finding generously covered at high water springs may be only covered to a dangerously modest extent at high water neaps.

DRYING OUT

Bilge-keelers, boats with lifting keels and flattish bottoms, or traditional long-keel yachts equipped with legs, all have access to a mouth-watering extra selection of overnight anchorages. Boats that can easily and safely take the ground as the tide falls away have a gold card pass to some magical corners of Brittany that deeper-draught boats will simply never bother or be able to visit. Havre de Rothéneuf is a prime example, a few miles northeast of St Malo. This spacious drying inlet is a glorious hideaway for boats that can creep in near high water springs and sit safely on the golden sand that starts to uncover once the ebb is running well out through the narrow entrance. Anyone who pokes their bows into Rothéneuf will find a charmingly enclosed slice of holiday coast that only a comparative handful of British yachts have experienced or will ever experience. The same applies to the various drying anchorages southwest of Ile Grande, just a mile or two north of Trébeurden marina, or the drying flats between Carantec and Ile de Callot where few visiting boats ever venture. At the far west end of the north coast, Portz Malo is a truly secret gem for bilge-keelers, as is Billiers on the south coast, near the mouth of the Vilaine.

Drying anchorages require a certain strategic judgment about the weather, because once you've settled down in some wild deserted spot, there's no escape until the flood comes back to lift you off. You therefore need either to have chosen an anchorage that's completely snug whatever the weather, or else made a sound decision about likely weather conditions for at least a tide ahead. You also need to use the largest-scale charts available for the best information possible on the state of the bottom before you position yourself for drying out. There are certain sections of the Brittany coast for which no large-scale Admiralty charts are published, so the large-scale French SHOM charts are the answer for dedicated dryers-out.

ESCAPING AT NIGHT

When staying overnight in any coastal anchorage, you should always have in mind some kind of plan about what to do if weather conditions should change unexpectedly in the small hours and threaten your security with a dangerous wind shift. Some anchorages are completely snug in any weather, in which case the only precautions needed are to make sure you have plenty of room to swing in any direction at all states of tide, and also that you have sufficient chain veered for the likely depth at high water.

But with anchorages that are not completely enclosed, you should work out, before turning in, how you'd escape safely in pitch darkness should the need arise. GPS and radar have simplified this problem for many boats in certain situations, although Brittany has plenty of anchorages that are intricately hemmed in either by low-lying rocks or reefs that just cover at high water. In such cases, close quarters pilotage is risky by GPS alone, while radar can only assist when rocks are high and large enough to provide strong, unambiguous echoes.

Bearings of nearby lighthouses, buoys or beacons can often give you a safe way out into clear water, or it may be possible even in the dark to skirt close to one side of a bay by eye where the coast is steep to and clear of dangers. Sometimes you can follow a charted depth contour with the echo-sounder, having made due allowance for the rise of tide at the time you'd be making your escape. In normal summer cruising you can often experience onshore wind shifts that are uncomfortable rather than dangerous, in which case it may be best to stay put until it's light enough to leave the anchorage safely. Shifting offshore and onshore breezes are particularly common down in South Brittany in hot summer weather. Certain anchorages along the Biscay coast are susceptible to a devilish *vent solaire* that can spring up just before dawn and freshen quite quickly to cause discomfort and agitation in bays or inlets that looked placidly idyllic the evening before in those magical dusky hours just after dinner.

Certain Brittany anchorages are difficult and some almost impossible to leave safely in the dark, so you should only stay overnight in such hideaways if you are pretty sure that the weather is settled and extremely unlikely to catch you napping. I have indicated these anchorages in the text as far as possible, although in the end every skipper has to make up his or her own mind about which spots are safe to lie in with that particular boat in the conditions prevailing at the time.

ANCHORS AND GROUND TACKLE

For countless boats of all types, the mechanics of anchoring have been greatly simplified by efficient electric winches and self-stowing anchors. A good winch also lets you carry anchors and chain genuinely heavy enough for your boat. You can find tables of recommended anchor weights and chain gauges in manufacturers' catalogues, though I have always added 25% to standard recommended weights. The peace of mind when the weather hardens is well worth the extra physical effort and initial expense.

I only use chain for anchoring, rather than chain and rope, because the deadweight on the seabed and damping effect in the hanging scope makes for more secure holding and minimum snatching. However, we do carry 50m of nylon to add to the chain if necessary.

Out of long habit and ingrained conservatism my favourite anchor is still a beefy drop-forged CQR. Aboard Stormalong we also carry a sizeable Fisherman's anchor, probably as old as the boat, which gives secure holding on weedy bottoms. The CQR was devised in the 1930s by Sir Geoffrey Ingram Taylor, an English mathematician who sailed from the muddy Thames Estuary. His strongly built articulated plough became the anchor of choice for most cruising yachts until new designs came along in the 1970s and 80s – first the claw-shaped Bruce anchor and later the 'self-launching' Delta.

The Bruce remains popular and sits neatly in a stemhead roller, but while plenty of boats use Bruce anchors around Brittany, they can be unreliable in thick weed. The 1980s Delta was essentially a one-piece advance on the CQR, a fixed-shank plough with a more complex shape and weight balance intended to make the anchor set more quickly.

Danforths are still popular as second anchors because they are relatively easy to stow in lockers, but as main anchors Danforths are my least favourite. They rarely penetrate weed satisfactorily and are apt to slide out of sand unexpectedly. Once a Danforth starts dragging it generally stays on the move and there's usually nothing for it but to haul up and re-anchor with more chain out.

Many well-travelled boats now use Rocna anchors, developed by New Zealand yachtsman Peter Smith after thousands of world cruising miles. The Rocna is essentially plough-shaped with a strong hoop welded around the flukes to ensure the anchor always falls in the best position for digging in. Friends of ours who sail in the

Caribbean say their Rocna has never failed to set first time, often within a metre of the dropping position – a measurement they have checked by snorkelling in the clear warm water. Another friend who has cruised for years around the west coast of Scotland is also a Rocna convert. His Arcona 40 carries a 20kg Rocna main anchor, with 40m of 10mm calibrated chain. The anchor has always set first time, except when dragging across smooth rock.

Whichever kind of anchor you use, the key to reliable anchoring lies in veering a generous scope of chain, so the pulling angle of the anchor is as near parallel to the seabed as possible. For anchoring overnight you should let out at least five times the maximum depth of water at high tide, which in practice means carrying much more chain than is normally supplied with a new 'ready-to-sail-away' boat. Suppose, for example, you wished to anchor in a relatively shallow corner of the North Brittany coast that had, say, eight feet of water at low water springs – which might apply to one of the more sheltered spots in the Anse de Paimpol. The spring rise of tide in this area can be around 32ft, so with a high tide depth of 40ft you ought to veer 200ft of chain for secure overnight anchoring. This sounds a lot, but in fact only represents six boat lengths for a 35ft boat. For anchoring in a low tide depth of 20ft in the same area, you'd ideally need to let out 260ft of chain (i.e. about 7½ boat lengths for a 35ft boat).

I always use an anchor buoy in Brittany, because you never know when your hook might get snagged in a rocky crevice. But be careful to use a buoy that doesn't look too tempting as a mooring. Having anchored off Camaret one summer night, we woke to find a sizeable French yacht neatly secured to our anchor buoy! Luckily it had been a quiet night. I rowed over to explain the situation and perhaps ask the skipper for some mooring dues, and he most kindly invited me on board for a cup of good French coffee and a warm buttery croissant.

Anchored off Port Manec'h in the Aven River

The River Rance up near sleepy Plouër marina

CHAPTER 1

CAP DE LA HAGUE TO ANSE DE PAIMPOL

The pronounced bight in the north coast of France between Cap de la Hague and the Anse de Paimpol is often known as the Gulf of St Malo. This is the fascinating and popular summer cruising ground that contains the Channel Islands and Iles Chausey, although I don't cover any of the Channel Island anchorages in this book as they would run to quite a sizeable volume of their own. Also, I think it's true to say that the anchorages of the two enclosing mainland stretches of Normandy and Brittany are generally less well known than those of the Channel Islands area.

Although the Gulf of St Malo is renowned for its powerful tides and copious rocks, you soon get used to passage planning with this in mind. Fast tides, in themselves, cause no great problems; they can, indeed, be turned to considerable advantage, making for surprisingly fast passage times if you judge your departures and arrivals carefully. GPS now eases the stress of navigational uncertainty for most yachtsmen, so that the main hazard caused by fast tides is their effect on sea state locally. Certain parts of this area are notorious for overfalls, especially with a weather-going stream and at springs. This is particularly true of the northern gateway to these waters, the Alderney Race.

Even a moderate wind over the tide creates nasty overfalls in the Alderney Race, so it's wise to go through as near slack water as possible. Ideally, yachts bound south should arrive at a position about two miles northwest of Cap de la Hague half an hour before HW Dover, i.e. about 4½ hours after HW St Helier. It is then slack water in the Race, but the southwest stream will be just about to start in your favour. Yachts coming north to leave the area face a trickier problem of timing, since it's important to carry the fair north-going tide up through the Race but to clear Cap de la Hague by an hour before HW Dover, i.e. four hours after HW St Helier.

Passages under sail within and across the Gulf of St Malo can usually be worked in six-hour stretches i.e. taking a full tide per leg. So that, for example, you can reckon about a tide each to sail from Cherbourg to Guernsey, from Guernsey to Jersey, from Jersey to

Le Grand Jardin lighthouse off St Malo

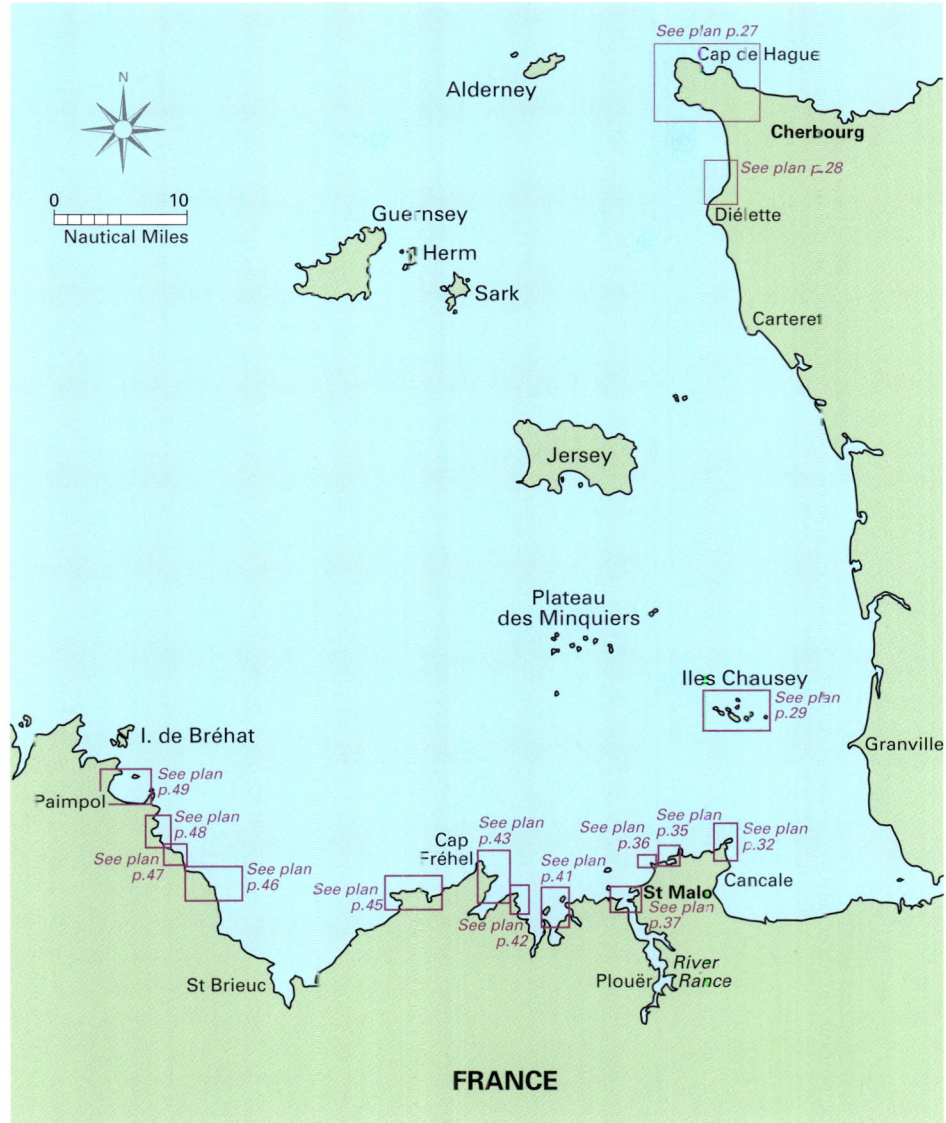

Granville or St Malo, from St Malo to Paimpol, and so on. In this respect, fast motor boats have more flexibility about passage timing, although it's generally even more important that they avoid heavy overfalls and hence time their arrival in potentially turbulent areas for as near slack water as possible.

While the Alderney Race is the most infamous tidal sluice in the Gulf of St Malo, there are several other 'gates' through which you need to time your passage carefully. The Swinge, between Alderney and Burhou, should always be treated with respect, because a wind over tide here – especially a southwesterly or westerly against a southwest-going stream – can give a yacht a rough ride from which there's no turning back. Sailing southwest from Alderney, it's best to leave Braye harbour about 2½ hours after HW St Helier to catch slack water in the Swinge. Yachts approaching the Swinge from the direction of Guernsey

THE COTENTIN COAST – BACK DOOR TO BRITTANY

The long west coast of the Cotentin peninsula is still a fairly well-kept secret for most British boat-owners. Many Channel Islanders know the three main yacht harbours well – Diélette, Carteret and Granville – but cruising yachts or motor boats coming down from England usually hop on the tidal escalator of the Alderney Race, make for St Peter Port and then look south towards Jersey, St Malo or Lézardrieux.

Of course Cotentin is part of Normandy, not Brittany, so in a way is slightly outside the scope of a book on Brittany anchorages. Yet the pronounced jutting arm of the Cotentin peninsula has a marked effect on cruising conditions in the whole Gulf of St Malo. Being so nicely hemmed around by mainland and islands, this splendid bay between Cap de la Hague and Ile de Bréhat provides a cruising area so diverse that it can keep you fascinated for many summers. The southern part of this area, sheltered between Jersey, Cotentin and Brittany, is blessed with a locally mild climate and generally quieter seas than you experience out in more open Channel waters.

In suitable weather, the Cotentin coast can also provide an interesting route down to North Brittany for boats based in the Solent area or further up-Channel. The first two anchorages in this book, Anse de St Martin and Anse Pivette, lie conveniently each side of Cap de la Hague for timing a short passage around this headland at slack water. In quiet summer conditions or a spell of easterlies, it's fascinating to follow the sandy Cotentin coast south from Anse Pivette into the Gulf of St Malo, keeping close in where depths and rise of tide allow. Within a couple of miles of this rather wild but intriguing shore, the streams are relatively weak for the Channel Islands area and it's not too hard to push a foul tide if you need to. However, you have to follow the charts carefully because this whole long littoral has many coastal shoals that either need a wide berth or have to be crossed above half-tide.

The attraction of this 'back-door' route towards Chausey and St Malo is a completely different atmosphere than you experience down on the rocky North Brittany coast. The first stretch to Cap de Flamanville is fringed by miles of superb sandy beaches that curve around the Anse de Vauville towards Diélette. We have anchored off this long beach in summer easterlies, particularly in the southeast corner of Anse de Vauville a short mile north of Siouville church spire. During a calm spell you can lie overnight here, although even in easterlies a slight rolling scend always seems to come in around high water.

Between Flamanville and Carteret you can keep fairly close in past a line of fabulous sand dunes, although the Roches du Rit jut out for half a mile offshore as you get down towards Cap Carteret. South of Carteret there are various drying banks and shoals between the mainland and Les Ecrehou. However, these patches are all well covered above half-tide when you can follow a more or less direct line between Cap Carteret and a waypoint midway between Basse Jourdan E Cardinal and La Basse du Sénéquet W-cardinal buoys. This gateway leads you south past the rocky plateau of Le Sénéquet that extends three miles from the low coast of sand dunes, guarded by an austere looking light-tower. For about seven miles south of Le Sénéquet the coastal shoals dry out for a good three miles from the low sandy shore. There are also numerous shoal patches in the offing, but they are all well covered above half-tide, when it's easy enough to thread the four-mile gap between the mainland and the east edge of the Iles Chausey plateau. Following this Cotentin route south from Anse Pivette you can feel the distinctive remote atmosphere of the west coast of Cotentin and you'll meet fewer yachts than you would along the more beaten tracks between the Channel Islands and St Malo.

will probably have carried the north-going stream up from the Little Russel and should aim to arrive opposite Burhou by 2½ hours after HW St Helier.

The Little Russel is the relatively narrow channel between Guernsey and Herm. The streams are strong in the Little Russel, especially at its north end, reaching six knots at springs near Roustel beacon-tower. With the uneven sea bed, a weather-going tide kicks up some nasty overfalls locally. Some of the most malevolent conditions can be experienced on the north-going stream in a stiff two-reef northeasterly. Many homeward-bound crews have found themselves hankering after the tranquillity of St Peter Port, having left with bravado only half an hour before, as they are swept through steep breaking seas at the north end of the Little Russel.

Down on the Brittany mainland there are powerful streams around the southwest corner of the Gulf of St Malo, especially near Plateau des Roches Douvres and in the outer approaches to the Lézardrieux River between Les Héaux lighthouse and the dangers seaward of Ile de Bréhat. In the Baie de St Brieuc the streams are more moderate but then, further east, you get a strong

Rocks and fort off St Malo

Chausey Sound

The anchorage off Clairefontaine

Flamanville west-cardinal buoy

sweep across the outer approaches to St Malo itself. In the southeast corner of the gulf, the Baie de Mont St Michel is a legendary area of strong tides, especially on a spring flood. Yachts are advised to stay well out of this potentially treacherous bay, except for the anchorages on the northwest side near Cancale.

The east side of the Gulf of St Malo, the Cotentin coast between Cap de la Hague and Mont St Michel, is actually in Normandy not Brittany, but I have included it in the book because it forms an integral part of the North Brittany cruising area. This can be a rather tricky and inhospitable coast, with much of it quite shallow for several miles offshore. The inshore streams are more moderate than elsewhere in the Gulf, until you get up near Cap de la Hague and the Alderney Race. There are good marinas at Diélette, Carteret and Granville, but this long west-facing shore is not really anchoring country, except for the few spots I have mentioned that can be used when the wind is either very light or settled in the east.

I have included a couple of fine-weather anchorages near Cap de la Hague – one to the south and one to the east – which can be useful while waiting for a fair tide, or simply to escape from the throng in high season. Anse Pivette, just east of Nez de Joborg, lies right on the boundary of a prohibited anchorage area opposite the nearby Atomic Energy Centre. Although this exclusion zone and the reason for it may seem rather forbidding to passing travellers, I've had no problems anchoring in this delightful bay, from which a path winds up the cliffs to a small restaurant. A few miles further down the coast, in quiet or easterly weather, you can anchor off the mouth-watering beach fringing Anse de Vauville, especially at the south end off the small town of Clairfontaine.

ANSE DE ST MARTIN

A couple of miles east of Cap de la Hague, this bay offers a possible anchorage for anyone awaiting a fair tide through the Alderney Race or, in settled weather, for a secluded overnight stop. St Martin offers reasonable holding in sand and is sheltered from west through south to southeast. Once the wind nudges north of west, though, an uneasy scend starts rolling in.

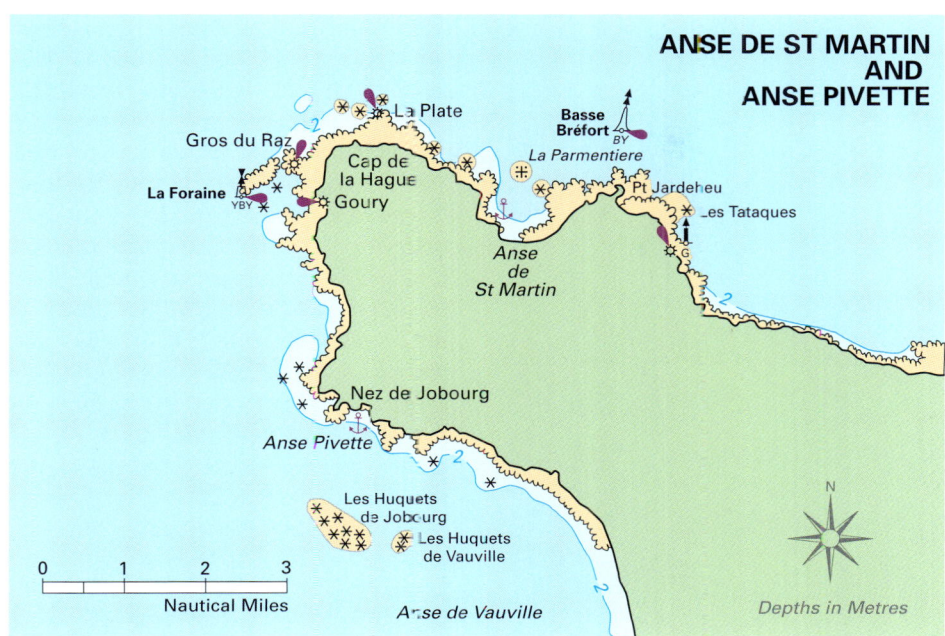

In westerlies, fetch up about 1½ cables east of Port Racine breakwater. On the approach, pick out Pointe du Nez – a low promontory at the west end of the sandy beach at the head of the bay – and keep its northeast tip bearing 187° fairly carefully. This line leaves La Parmentière (awash at LAT) 1½ cables to the east and Les Herbeuses (above-water rocks 3–10m high) about two cables to the west.

ANSE PIVETTE

Lying close east of Nez de Jobourg, this tiny bay offers pretty fair shelter in winds from between due north and east, clear of the powerful streams through the Alderney Race. Approach from a position three-quarters of a mile southwest of Nez de Jobourg, preferably at slack water, steering at first to leave Les Calenfriers (dries 1.1m) and Les Huquets de Jobourg (dries 5m) well to starboard (i.e. to the south).

Having skirted La Ronde, a tail of drying rocks extending nearly a quarter of a mile south of the Nez, enter the bay from just east of south. Tuck in close under the Nez, fetching up in about 3–4m.

CLAIREFONTAINE

The seaside village of Clairefontaine lies about a mile northeast of the entrance to Diélette harbour, just beyond a wide expanse of rocky coastal ledges that dry for up to four cables offshore. The village is right at the south end of a fabulous four-mile stretch of golden beach and dunes that fringe the long Anse de Vauville. In settled easterly weather I have anchored comfortably off the beach at Clairefontaine, 3–4 cables out from the charted coastline (the sand dries a long way out) and a couple of cables north of the edge of the drying coastal ledges that straggle north from Diélette.

This is a spot for languid summer weather when the pressure is high and the wind is either negligible or definitely offshore. Diélette is very close if the

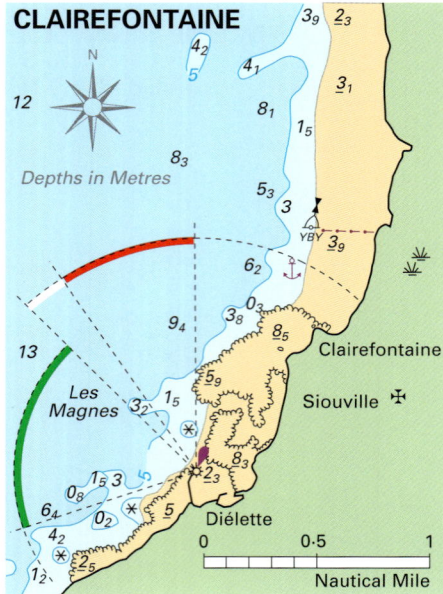

weather should change, although you need at least two hours' rise of tide to get into the outer harbour.

ILES CHAUSEY

The timeless Chausey islands lie down in the sheltered crook of the Gulf of St Malo, some 20 miles southeast of St Helier, half a dozen miles WNW of Granville and about 16 miles northeast of St Malo. The plateau is compact and practically steep-to – almost a perfect oval on the chart – stretching about six miles from east to west and 2½ miles north to south. On all sides except the far east, you can edge in safely towards the outer reefs until you pick up your bearings from Admiralty Chart 3656 or, better still, the French SHOM Chart 7134.

Chausey has several rather special anchorages. The most popular is the main harbour area of Chausey Sound, but in quiet weather you can anchor two cables west of the south entrance to the Sound, in the bay known as Port Marie.

More adventurous explorers might try the secluded hideaways a couple of miles ENE of Grand Ile, more or less in the middle of the plateau.

Chausey Sound This narrow but easily navigable channel cuts between Grande Ile Chausey and the main mass of drying reefs and islets to the east and north. The sound provides good shelter from all winds at LW, but becomes uncomfortable near HW in any fresh weather from the northwest or southeast. Streams run fast through the sound around half-tide, so a certain amount of care is needed when planning trips ashore in the dinghy. Rowing ashore at peak tides is dangerous and you certainly wouldn't want to fall in.

Entry to Chausey Sound is straightforward from the south, between three E Cardinal beacons off Pointe de la Tour and a green conical buoy on the east side of the fairway. The more delicate north entrance requires careful pilotage and is navigable from about 2½ hours before to 1½ hours after HW.

There are now many moorings in Chausey Sound, including several sturdy visitors' moorings in the reach opposite the Grande Ile landing slip. On summer weekends these buoys are often packed with yachts rafted up to six abreast, sometimes more. Anchoring is still just about possible in various gaps between moorings, although you should always buoy your anchor against the risk of catching a mooring chain.

The northwest part of Chausey Sound has the most free space for lying to your own hook, although the upper reaches dry at LAT and you'll find greater scope for anchoring up here at neaps. You can also anchor in the southern part of the sound near the entrance, about 200m south of La Crabière-Est S Cardinal beacon towards the *west* side of the channel.

CAP DE LA HAGUE TO ANSE DE PAIMPOL

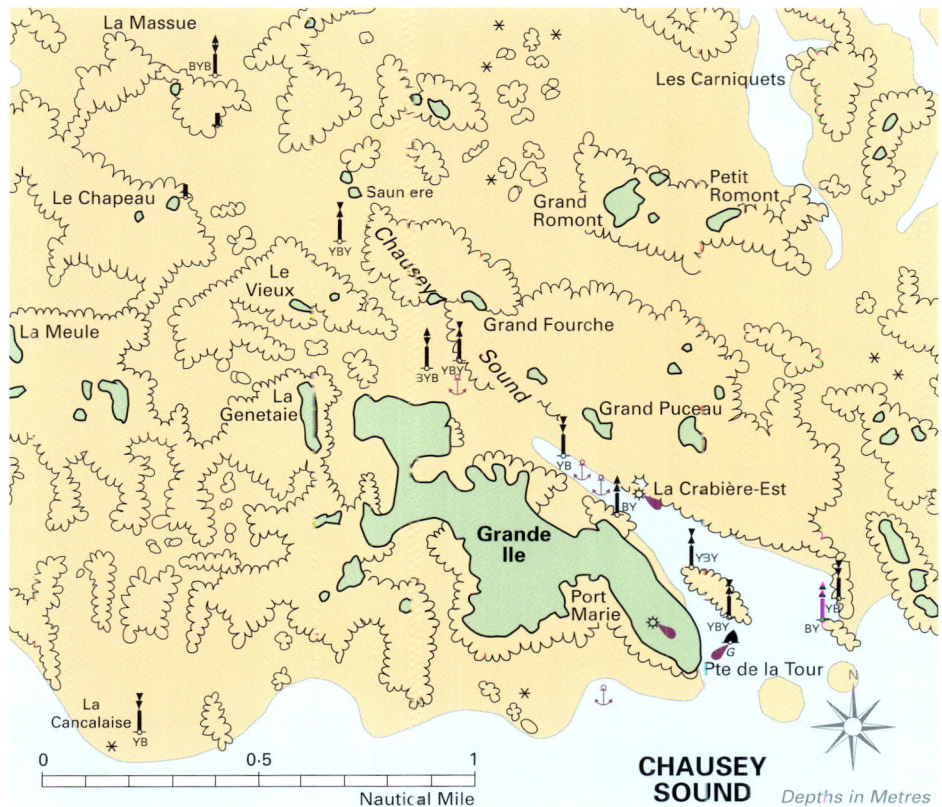

Chenal Beauchamp This deep and mostly fairly wide channel cuts right through the eastern half of the Chausey plateau, dividing the banks, reefs and islets into two distinct sections. The northern part of this sound runs north-west to south-east and the southern part turns more or less north-south. In quiet weather you can easily enter the southern half of Chenal Beauchamp from southward and anchor just over a mile in, towards the north end of this section in about 3m datum depth.

This southern entrance is actually simplest towards low water, when the drying edges of the channel are clearly defined. Approach the plateau from due south, making for Tournioure isolated danger (BRB) spar beacon just west of La Chapelle above-water rock, which

SECRET ANCHORAGES OF BRITTANY 29

Chausey Sound

itself stands about a third of a mile southwest of the white stone beacon known as La Tourelle des Huguenans. You can pass either side of the Tournioure spar, but the deepest water lies close on its west side.

Having skirted round Tournioure, make good just a smidgen east of north, leaving Iles des Huguenans to starboard and a largish outcrop of drying rock to port. A useful steering transit, if you can make it out, is to keep La Culassière rock (just over a mile ahead, four metres high) bearing 002° and nicely open to the west of the distant but prominent L'Etat beacon (white with a black top, standing on a hump-shaped islet).

Mouillage de Beauchamp Just beyond the main entrance gap between the Huguenans to starboard and the rocky outcrop to port, leave an E cardinal spar beacon off Roche Ango well over to port. Then continue broadly north into the channel but keep towards the east side of the deep water to avoid a shallow rocky patch (drying 1.7m) jutting out from the west side. You can anchor a cable NE of a second E cardinal spar (L'Herbier) in about 3m datum depth. Below half-tide this spot is sheltered by Chausey's vast acres of drying reefs, a fascinating daytime retreat in fine summer weather, especially around springs with an early afternoon low water.

Les Roquettes à l'Homme This intriguing anchorage lies about halfway along the Chenal Beauchamp, cocooned between the west and east halves of the Chausey plateau. You can approach Les Roquettes from either the north-west or south, although I think the southern approach is easier and certainly more convenient if you are visiting Chausey

The middle reaches of Chausey Sound

from nearby Granville marina. Aim to cast off from Granville a couple of hours after high water, well before the marina sill closes.

From Mouillage de Beauchamp you are almost there. From a cable north-east of L'Herbier E cardinal spar, follow the narrowing channel north-west towards a clear gateway between a N cardinal spar and a S cardinal tower. Anchor a cable beyond this gap in about 4m datum depth, NE of the rocky outcrop called Les Roquettes à l'Homme, whose peak dries 13m.

Near low water you are sheltered by the exposed expanses of the Chausey plateau. I have stayed here overnight in quiet weather, though you must veer plenty of chain for the high water. The eerie remoteness isn't everyone's cup of tea, yet Roquettes is an amazing place on a clear night with no other boats in sight. You'll see the single sweep of Grande Ile lighthouse, the four flashes of Granville's Pointe du Roc and the occulting Pierre d'Herpin to the south.

Pierre d'Herpin lighthouse

Grande Ile Chausey

Oyster barges at Cancale

CANCALE

Although Cancale harbour dries at half-ebb and its immediate approaches dry at low springs south of Pointe de la Chaîne, there are several possible anchorages off the coast between Pointe de la Chaîne and Pointe du Grouin, depending on wind direction and the height of tide.

Pointe de la Chaîne Around dead neaps and up to halfway towards springs, boats of moderate draught can stay afloat 2–3 cables SSW of the green beacon off Pointe de la Chaîne. You can easily see the short tongue of comparatively deep water on Admiralty

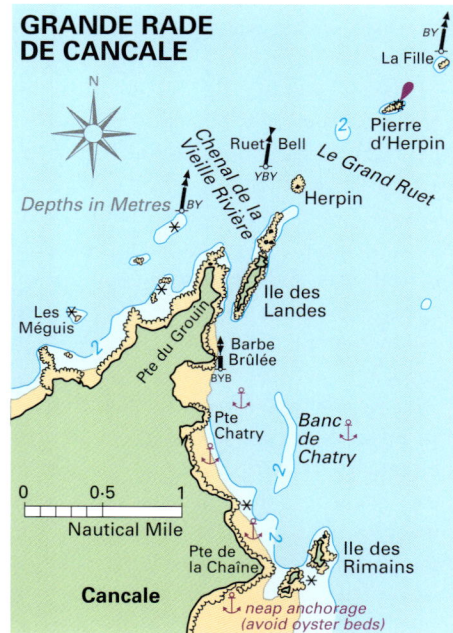

Chart 3659. This spot is snug in quiet or westerly weather, although if too much north comes into the wind a swell starts rolling down from Pointe du Grouin. You can land at the nearby beach, from where Cancale town is not far to walk.

Rade de Cancale There are three shallow bays between Pointe de la Chaîne and Pointe du Grouin, any of which can be used by keel boats in light to moderate winds from between northwest through west to south. These bays contain local moorings in the summer, but there's still room to anchor with reasonable holding in muddy sand. Tuck in as close as you can to avoid the worst of the tidal stream and any coastal roll. The approach is straightforward, either from the north via Le Grand Ruet or the Chenal de la Vieille Rivière, or directly into the Grande Rade de Cancale from the northeast. If the wind freshens from the northwest or even the west, an uncomfortable scend will start sneaking round Pointe du Grouin and find its way into these bays.

OYSTER BEDS OF CANCALE

Cancale is a veritable oyster-growing Mecca and the attractive harbour front is lined with seafood restaurants offering tantalising menus of *fruits de mer* and *dégustation des huîtres*. If you stand on Cancale jetty and look out across the Bay of Mont St Michel, oyster beds marked by withies seem to stretch for miles across this vast shallow expanse of drying muddy sand with its notoriously extravagant tides. At low water, tractors and trailers drive out across the sands, engaged in the mysterious business of tending, turning and moving millions of slowly developing oysters. Around high water, flat bottomed boats with wide flat decks carry baskets of maturing oysters hither and thither.

To the layman, the whole business of oyster cultivation looks something of a black art, but there are two distinct stages of production carried out at Cancale – maturing and cleaning. The beds well out in the bay are used for maturing and the oysters stay there for two years in mesh sacks, known as *poches*, which are supported above the mud in wooden racks. The oyster 'guardians' go out frequently to turn the *poches* when the tide is out. Later, the oysters are moved to the cleaning *parcs* (or *dégorgeoirs*) just below Cancale harbour, where they gradually flush themselves with mud-free water in an area where the relatively slack tidal stream picks up less silt. The mud is cleaned off the shells by passing the oysters along a conveyor through a jet of water. It's a long, patient process and most Cancale oysters are four or five years old when they appear on your plate. The climax to this slow maturation is almost shockingly quick, as you squeeze a little lemon juice on your opened oyster and slither it down live in a very few seconds. But as you finish your dozen and look ruefully at the empty shells, you can be sure of one thing – there are plenty more oysters coming along behind.

Oyster operations at Cancale

LEGENDS OF MONT ST MICHEL

Surrounded by miles of empty drying sands covered twice a day by the biggest tides in Europe, the famous Mont St-Michel is rarely approached by boat and always looks strangely unreal and enigmatic from offshore. In 708 AD 'the Mont' was just an isolated rock, Mont Tombe, which provided an occasional refuge for hermits or locals cut off by the tide. However the Bishop of Avranches, Aubert, is said to have had a dream in which the Archangel Michael appeared and told him to build a church on the rock. For some time the Bishop ignored this arduous summons, trying to convince himself that it wasn't really St Michael who had delivered this directive. He underestimated the Saint's persistence and the dream apparently recurred night after night. It seems that the message was finally brought home to Bishop Aubert when the angel put his finger above the Bishop's right ear, making a hole right through his skull.

Preparations for building an oratory on the rock then got under way. (I suppose the impetus for numerous large building contracts have had less credible starts that this.) Standing right on the border between Normandy and Brittany and with its natural defences of tide and quicksands, the small island of Tombe became an important site during the Middle Ages. In 966 AD Benedictine monks were installed on the island and the enlargement of the abbey began. Mont St Michel, at it was then called, soon became a site of pilgrimage but also a garrison that was heavily defended during the Hundred Years War.

By the time of the French Revolution the abbey, because of its remoteness, was used as a prison, and it continued as such until 1863. The present guided tours of the abbey bring dramatically to life its various uses down through the ages. Because of its gradual enlargement and development over many centuries, it's easy to become disorientated and lost in the many passages and rooms inside the abbey.

Below the abbey, shops and restaurants wind up the tiny streets that are packed with tourists during the summer. Mont St Michel is the most popular attraction in France outside Paris, so there's something to be said for just glimpsing its fairytale profile from some distance offshore. On passage between Granville and Cancale you can usually see the Mont in the corner of the bay, shimmering on the horizon like a mirage.

A local delicacy of the area is *l'agneau de pré-salé*, the delicious tender lamb that has grazed on the thin grassland of the marshes around the Baie du Mont St Michel. The subtle flavour of this meat comes through like a delicate natural seasoning, with almost a tang of sea air. A fairly dry but fruity rosé wine, slightly but not over chilled, goes well with *l'agneau de pré-salé*.

HAVRE DE ROTHENEUF

The narrow entrance to this large natural inlet lies midway between Pointe du Meinga and Pointe de la Varde. Rothéneuf dries, but there are various sandy beaches where suitable boats can take the ground on bilge-keels or legs in perfect shelter. One of the best spots is in the southwest corner of the inlet, close inshore with Rothéneuf church spire bearing about WSW.

Approach Havre de Rothéneuf near HW from a position half a mile south of Rochefort W Cardinal beacon tower and bring the edge of the east head of the entrance to bear 163° (with Rochefort bearing 343° astern). Follow this line and pass close east of the green beacon in the middle of the entrance. Patches of drying rocks lie either side of the approach, so allow carefully for any cross-tide.

POINTE DE LA VARDE

There's a tiny daytime anchorage just east of Pointe de la Varde, sheltered in winds with any south in them. You simply follow the charted marks for the Chenal de la Bigne and turn off opposite Pointe de la Varde, making the final approach from the northwest to pass close to the northeast tip of the point. An overnight stop here is not recommended unless the weather is very settled, because it's almost impossible to find a safe way out in the dark if you have to.

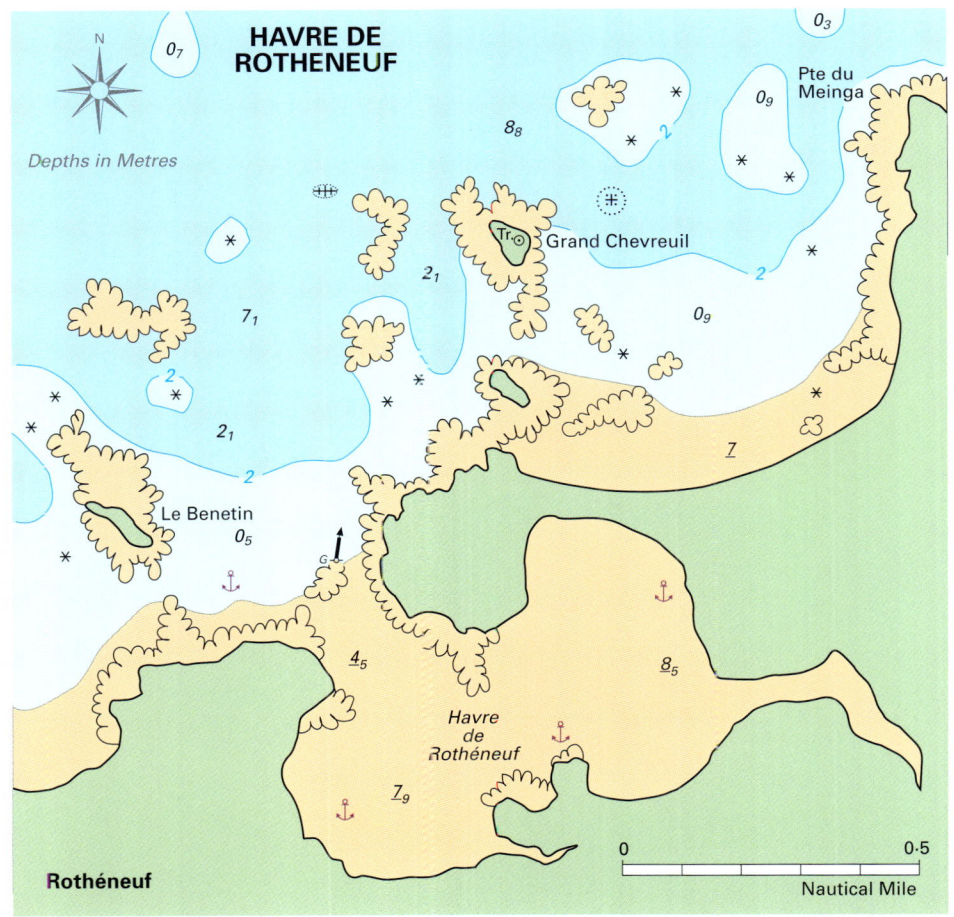

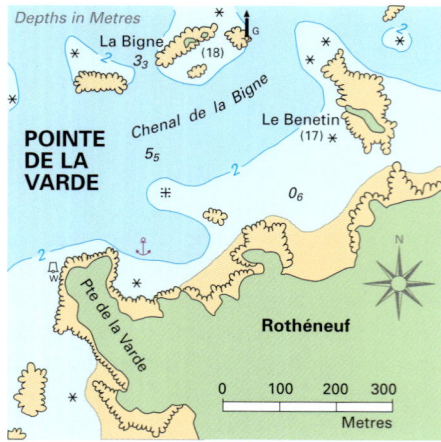

ILE DE CEZEMBRE

This distinctively-shaped island guards the approaches to St Malo and is usually the first part of the estuary you can identify when approaching from Guernsey or Jersey west-about Les Minquiers. Le Grand Jardin lighthouse stands not quite half a mile southwest of Cézembre, the key mark for the two main St Malo entrance channels – La Grande Porte and La Petite Porte.

In quiet summer weather, you can anchor for the day close south of Ile de Cézembre, a memorable spot for a lazy lunch especially as the tide ebbs away. A period midway between neaps and springs is good for Cézembre, with a moderate low water in the early afternoon. The tide will not be too low, which allows you to anchor reasonably close to the island, but it should be low enough that the drying rocks straggling SSE from Cézembre, and also those around Le Grand Jardin, are well enough exposed to help keep out any swell.

Half-ebb is a good time to arrive and Cézembre is easy to approach from St

St Malo main breakwater

CAP DE LA HAGUE TO ANSE DE PAIMPOL

Malo by first following the well-marked Rade out towards Le Grand Jardin. When No 10 red buoy is abeam to starboard, turn north towards the prominent old buildings near the centre of Cézembre, keeping them bearing just a shade west of due north true. This line passes clear between Les Bonnes Hommes rocks (awash at datum) to port and Les Herbiers (drying up to 4.5m) to starboard. If Les Herbiers are showing, you can fetch up with the highest part bearing due east about a cable off, or edge in closer towards Cézembre as the depth allows.

RANCE ESTUARY

In the past there have been two possible anchorages in the upper estuary between St Malo and the Rance Barrage – off St Servan in Solidor Bay on the east side, or off the wide Anse de Dinard on the west side. In both these locations, local moorings have steadily encroached into the space available for lying to your own hook, but there's still just about room for those who prefer to trust their own ground tackle rather than the unknown strength of a borrowed mooring. The best spot in Solidor Bay is about a cable due west of Pointe des Corbières, inside the Solidor Bank. The Anse de Dinard

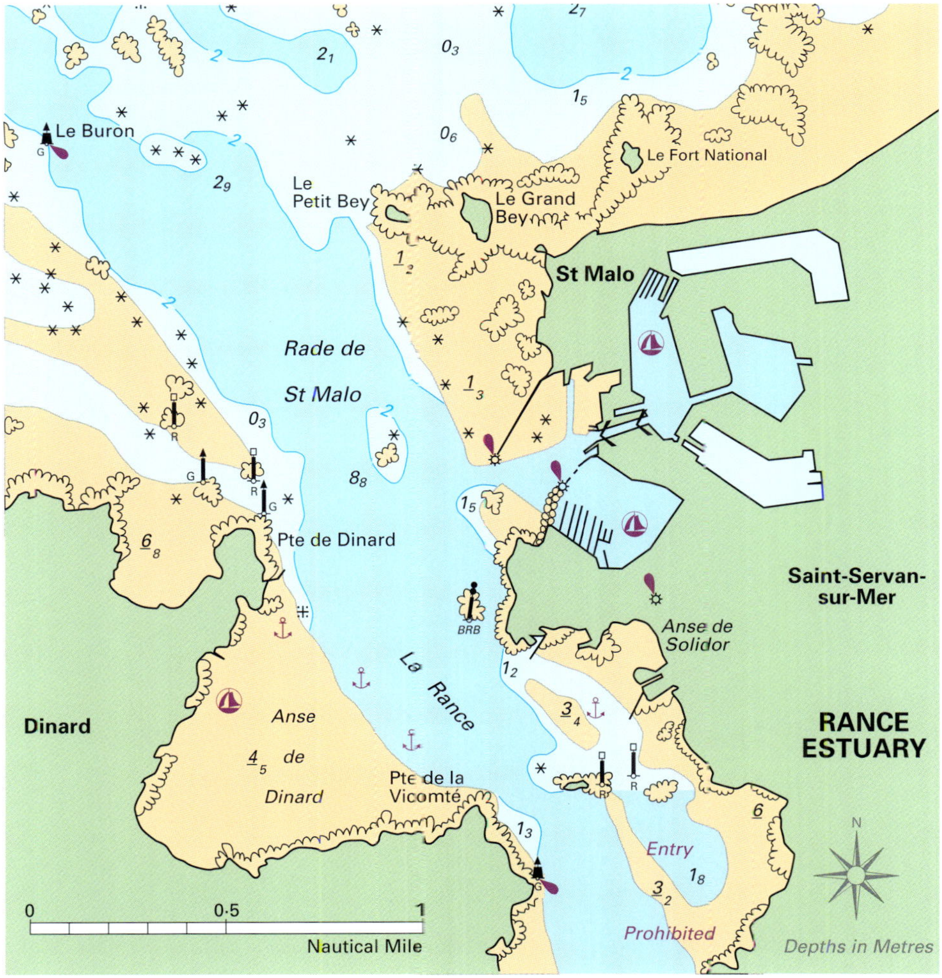

SECRET ANCHORAGES OF BRITTANY 37

Surcouf

Duguay-Trouin

ST MALO PRIVATEERS

The origin of this unique town was a remote 6th-century monastery. St Malo himself was Bishop of Aleth, the ancient settlement that has since become St Servan. Some of the old walls of Aleth can still be seen on the peninsula above Port des Sablons.

In these early times, the rocky outcrop where St Malo now stands was surrounded by salt marshes and drying sands. A prosperous walled town developed on this outcrop and Aleth gradually declined. The Malouins became famous shipowners, merchants, explorers and privateers and for at least 200 years managed to sustain themselves as an independent republic, professing no allegiance either to France or Brittany, according to their famous code:

Ni Français, ni Bretons, Malouins seulement.
(Can't you just hear them saying it?)

In the early 18th century the sea broke through to leave St Malo as an island, which rather suited the privateers, but later a causeway was built to connect the walled town to what is now the 'seaside' area of Paramé. The docks and basins have gradually developed to the south.

The old town within the walls still feels like an island and you can imagine how it became such a useful stronghold for those swashbuckling Corsairs who wreaked such havoc on British shipping in the days of sail. The most successful Corsair captains such as Duguay-Trouin and Surcouf amassed huge fortunes, first by their legalised piracy and then by becoming respectable shipowners and merchants. These wealthy men built stately town houses in St Malo, and also established summer mansions, known as *malouinières*, further out in the country on the banks of the Rance.

It's interesting that St Malo served as a stronghold again over 200 years later, in the middle of the 20th century. Towards the end of the Second World War, in the weeks following D-Day, the struggle to liberate Brittany was hard and bloody. Units of the German high command retreated to St Malo and the American VIII Army Corps bombarded the town heavily. By late 1945 the old *cité intra-muros* was virtually a pile of rubble. Its rebuilding has been a triumph of authentic restoration.

CAP DE LA HAGUE TO ANSE DE PAIMPOL

St Suliac village on the River Rance

dries right out at springs, but close to neaps you can tuck into the north corner of the bay and still find a spot to anchor.

RIVER RANCE

Having passed through the Rance barrage lock into the river proper, you emerge into one of the most magnificent stretches of sheltered water anywhere in France. Although most of the pronounced river bays are now well filled with moorings, you can still find room to swing to your own anchor. Try to tuck as close inshore as possible to avoid the powerful river currents and bear in mind that the rise and fall is rapid when the barrage sluices are open for the tidal generators.

In the shallower upper reaches of the Rance, particularly between St Suliac and Mordreuc, keel boats need to pick their spot with care, sounding carefully before anchoring. One of the problems with the Rance is that you're never quite sure how far the electricity engineers will let the ebb run away before shutting their sluices, so you always need to allow plenty of depth in reserve when anchoring.

There's a useful deep-water pool just downstream from the two bridges at Port St Hubert, where you can stay afloat at any state of tide, including very low springs. Above the bridges at low springs, only a narrow shallow channel remains close around the west shore, when many of the local boats moored opposite Plouër go aground.

ST BRIAC-SUR-MER

Just five miles west of St Malo there are several fascinating indents in the North Brittany coast that few cruising yachts visit. The first leads in behind tiny Ile Agot towards the low-key resort of St Briac-sur-Mer, which is much smaller than Dinard but popular with French families for traditional summer holidays. The immediate approaches to St Briac dry, but in quiet weather you can anchor

TAMING THE RANCE

The Rance is probably my favourite Brittany river, meeting the sea above St Malo at that stark concrete barrage built in the 1960s to harness the power of the mammoth tides hereabouts. The Rance 'tides' above the barrage are largely controlled by the power station engineers, who open the turbine sluices according to a published schedule. Before the dam was completed, the Rance was quite a wild river with fast-flowing streams and a spectacular rise and fall all the way up to Le Châtelier. In the early 1960s, a car ferry shuttled across the lower estuary between St Servan and Dinard, a rather sporting crossing that relied a great deal on the flair of the ferry skippers. Now the traffic drives across the barrage, which looks grimly utilitarian as you approach and head for the lock at its west end. Wide expanses of estuary are buoyed off to prevent boats straying into the dangerous currents stirred up by the generator turbines. On a first visit, you wonder what might lie behind this somewhat hostile looking set-up.

But as the lock gates creak open and you poke your bows into the Rance, the peaceful river vistas create an instant sensation of well-being. These almost classical French scenes seem to whisper a soothing message that, whatever else happens in the hectic world outside, the Rance will always look and feel exactly like this.

The valley is rural but not wild or remote. Sleepy hamlets, neat farms and ancient woods placidly enclose a very civilised navigable corridor steeped in that unique French blend of privilege and *égalité*. You pass rambling houses owned by the same family for generations, mostly locked and shuttered except during *les grandes vacances* in July and August. Alongside the quiet wealth of these retreats, locals and visitors enjoy easy access to the river, which has plenty of public slips and landings where anyone can launch a small boat and savour all the Rance has to offer.

The Rance tides lag behind the estuary tides by up to three hours, with both ebb and flood concentrated into a three to four-hour period instead of the usual six. The important point for visiting boats is that the streams flow swiftly in the Rance when they *are* running, especially at springs; the level also rises and falls almost visibly during this compressed period. Remember this when anchoring or making dinghies fast to quays or slips.

The River Rance opposite Montmarin

and stay afloat off the shallow entrance, which is effectively the mouth of the Frémur River.

The best way to arrive from St Malo is by taking the Chenal du Décollé at high water. The entrance to this shallow channel lies just opposite the St Malo pierhead and Admiralty Chart 3659 shows the route clearly. With plenty of rise of tide, it's easy to follow the beacons westward out of the Rance estuary to the final gateway of red and green beacons off Pointe du Décollé. From here, with the tide well up, you can turn WSW to pass 250-300m south of Nerput E Cardinal beacon tower and then steer southwest to pass midway between Ile Agot and Pointe de la Haye.

Curving gradually south round Ile du Perron and its off-lying red spar beacon, you can then turn in east towards the Frémur River entrance and anchor a cable due south of this red beacon. If the tides are closer to neaps than to springs,

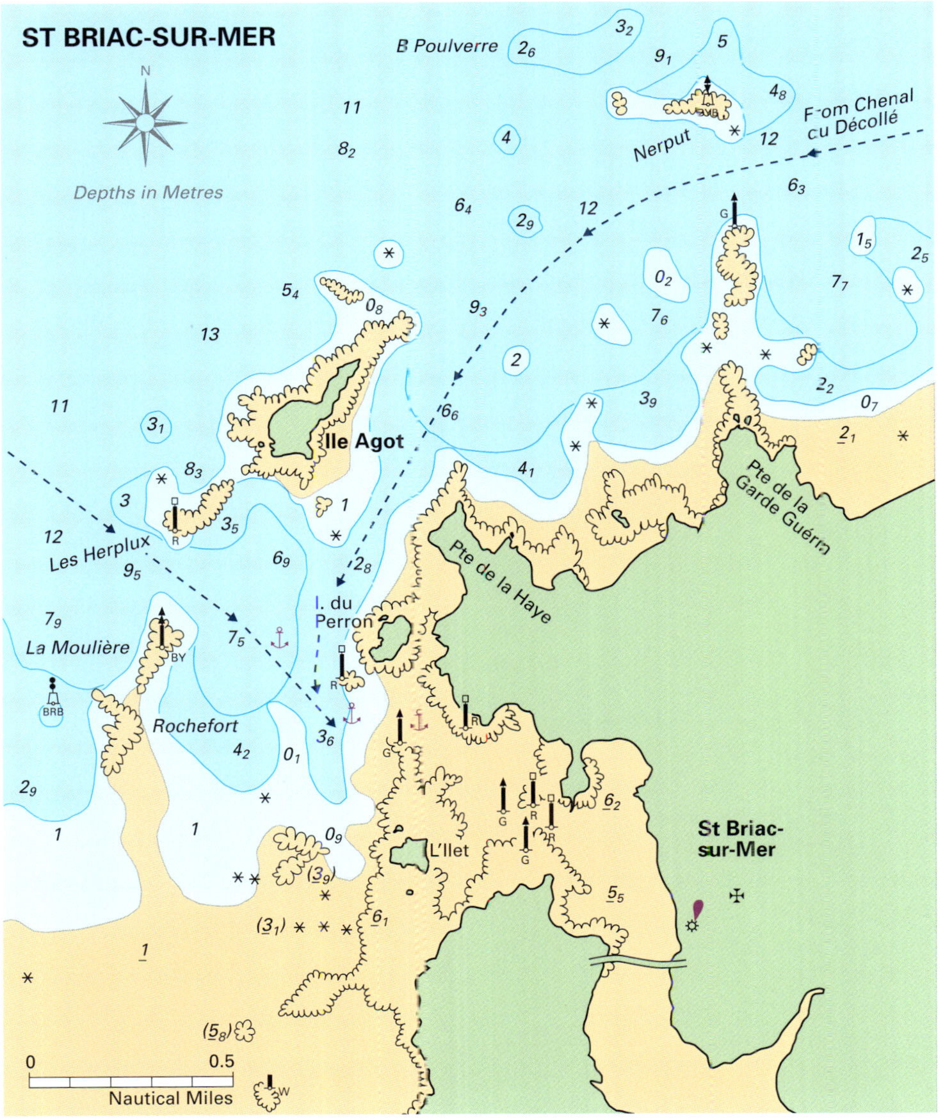

St Briac-sur-Mer

SECRET ANCHORAGES OF BRITTANY

you can edge a little further into the river and still stay afloat at low water.

As the ebb falls away, this spot becomes increasingly sheltered from any outer swell as the rocks and shoals between Ile Agot and Rochefort uncover. This anchorage is perfectly snug in offshore winds from due east through south to southwest, but it can also be very pleasant over the low water in a light summer northwesterly so long as there's no significant swell outside. In a more active northwesterly, the anchorage off Ile des Hébihens is more sheltered (see below).

ILE DES HEBIHENS

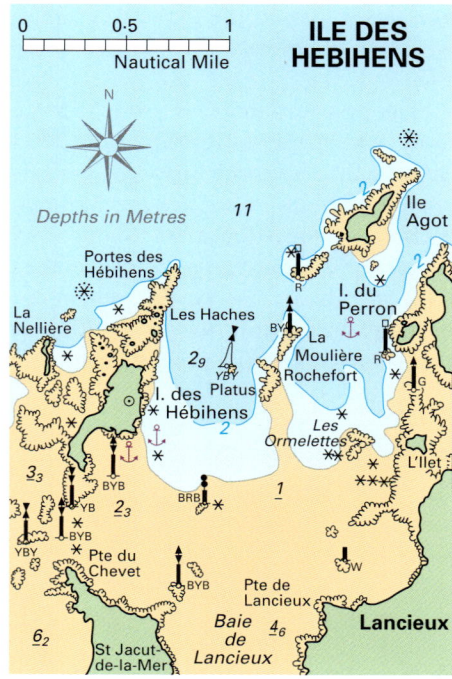

There's a delightful sheltered anchorage off the south side of this small private island, which lies a couple of miles west of St Briac-sur-Mer and 2½ miles southeast of Pointe de St Cast. Hébihens can be identified by its prominent tower (33m high). You can approach from due north at any state of tide, leaving Portes des Hébihens above-water rocks (9m high) about 200m to the west and then steering to leave more rocks and then the island itself a little further, maybe 300m, to the west. To port you should see the ruins of the old Platus beacon tower, now guarded by a small W Cardinal buoy which you leave to port. Admiralty Chart 3659 takes you in easily.

Watch the echo-sounder carefully on the way in, fetching up a little way into the pronounced bay (almost dries at LAT) formed by the south shore of the island and the narrow promontory that extends towards Pointe du Chevet on the mainland. There are numerous rocky dangers between the tip of this spur and Pointe du Chevet.

The bay is a charming place to linger on a warm summer day and the clear water over the fine sand is enticing for swimming. Neap tides allow you to anchor closer in, but near springs you have to stay further round towards the east side of the island – the charted depths here are a metre at LAT, but in practice there's usually enough depth to stay afloat.

You can reach Ile des Hébihens from St Malo by coming out through the Chenal du Décollé near high water. Follow the St Briac directions above to pass between Ile Agot and Pointe de la Haye, but then cut westwards between Les Herplux red beacon and La Moulière N Cardinal spar.

POINTE DE LA GARDE

In moderate westerly weather, there's a pleasant anchorage close southeast of Pointe de la Garde, a mile south of St Cast. La Garde has a clear approach from the northeast and the anchorage gives fair holding in sand with a depth of about 1.2m at datum. There's a landing slip on the southeast side of the point.

BAIE DE LA FRESNAIE

The wide entrance to this shallow bay lies two miles southeast of Cap Fréhel. Most of the bay dries and is taken up with mussel beds, but there's a shallow anchorage on the west side of its mouth, a couple of cables SSW of Pointe de la Cierge. On several occasions I have edged my way in here in quite murky visibility, first closing from the east with Pointe de la Latte and its easily identified fort, and then sounding into the bay close along the west shore. Anchor as close inshore as the depth allows. This spot is sheltered in winds from WNW through west to southwest, with good holding in sand and mud.

There's also an anchorage just south of Pointe de la Latte itself, used sometimes by French yachts and sheltered broadly from the west. I find this rather a forbidding spot, though, and there are often overfalls and uneasy tidal eddies off the headland.

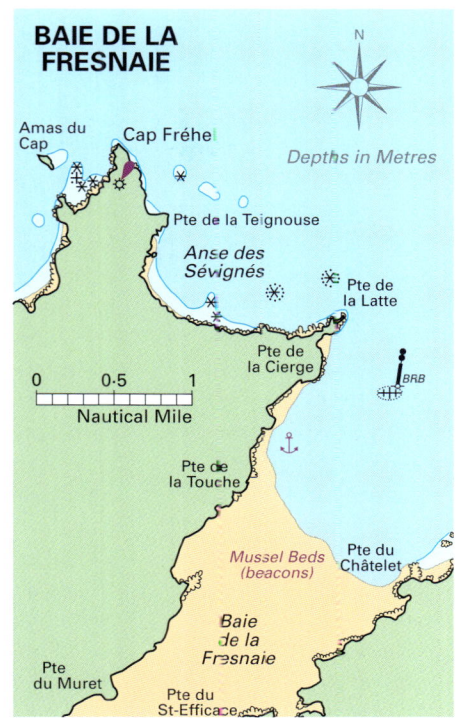

LES BOUCHES D'ERQUY

This off-beat anchorage makes an interesting diversion if you are on your way west towards Erquy or perhaps Binic, although it should only be approached in quiet weather or with the wind in the south. From Cap Fréhel, come onto a line for the Chenal d'Erquy, but turn off shorewards just before Plurien church spire bears due south true.

Head towards the long, sandy beach that fronts Sables d'Or-les-Pins, keeping Plurien spire bearing south. Leave Ile St Michel (with its small chapel) three-quarters of a mile away to starboard and steer to leave Rocher Bénard (a conical above-water rock close inshore) about a quarter of a mile to port. When Ile St Michel is abaft the beam, head southwest for the west end of the dunes and fetch up in about two metres LAT, with Point du Champ du Port bearing WSW about three cables off.

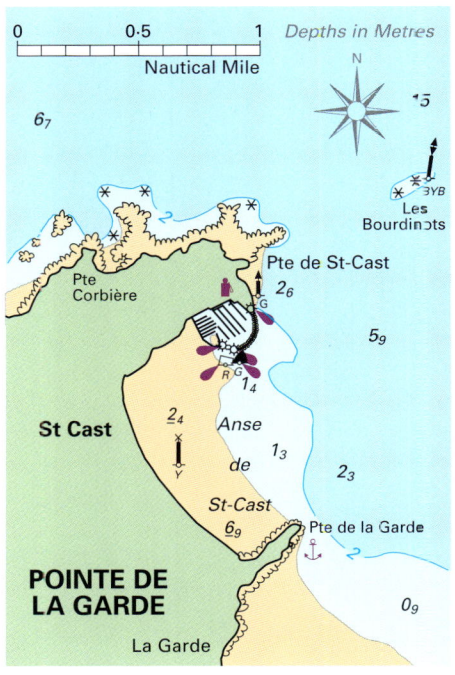

FORT DE LA LATTE

As you approach the Brittany coast to the west of Saint Malo, the main landfall feature is the striking granite headland of Cap Fréhel, which rises almost sheer from the sea for 75 metres. Topped by its powerful lighthouse, Fréhel is an unmistakable landmark, but less than two miles to the east, on a slightly lower headland, stands an imposing and rather ornate castle, Fort de la Latte. The castle has pretty good sea views, being perched right on the end of the jutting finger of Pointe de la Latte, the promontory that helps shelter Baie de la Fresnaie from the northwest.

This sturdy medieval fortress was the home of the Goyon-Matignon family during the 13th and 14th centuries. Its naturally defensive position on the headland, with crevasses crossed by drawbridges on the landward sides, made Fort de la Latte virtually impregnable throughout numerous wars. Vauban made some small modifications, but changed the castle little, which was saying something. In 1715 the Scottish 'Old Pretender', James Stuart, used La Latte as a base when he was gathering his forces before trying to take the English throne. In the Second World War the Germans used the fort as a lookout and defensive position.

In more recent, less warlike times, several films have been located there, probably the most famous being *The Vikings*, directed by Richard Fleischer. The cast included Kirk Douglas and Tony Curtis, so you can picture the extravagant scenes around the ramparts and drawbridges.

The fort was constructed as a rectangle with enormously strong towers around its walls and a traditional keep in the centre. It is open to visitors from Easter to September and a climb up the 15th-century lookout tower is included in the tour. Just outside the castle stands a rather enigmatic-looking megalith, said to be either the finger of the giant Gargantua or else a mark for the place where his head rests in its tomb. A largish chap, his feet are apparently 25 kilometres away at Saint-Suliac.

Fort de la Latte

CAP DE LA HAGUE TO ANSE DE PAIMPOL

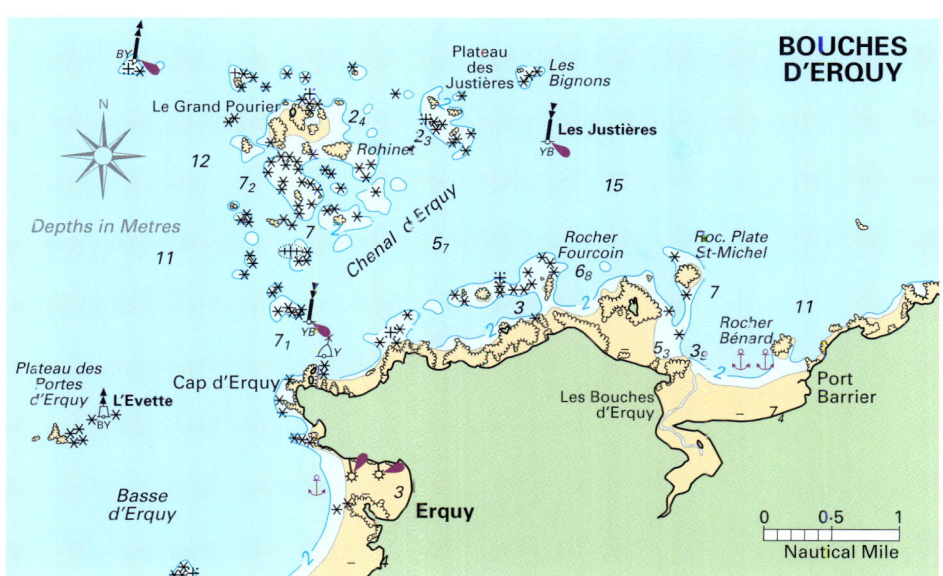

An alternative anchorage is off Fort Barrier, two cables SSW of Rocher Bénard. A good time to lie at either of these spots is midway between springs and neaps. You'll have a little more anchoring depth to play with than at full springs, but the rocks and islets north of Point du Champ du Port are still well exposed at low water to provide good shelter from swell.

RADE DE PORTRIEUX

This traditional roadstead anchorage almost comes under the 'harbour' category, but is worth bearing in mind as a reasonably secure place to swing to your own hook on the west side of the Baie de St Brieuc. The Rade lies SSE of the entrance to Port d'Armor marina and more or less opposite Portrieux old harbour, which dries. In quiet weather or offshore westerlies you can anchor

Rohein islet and west cardinal beacon

SECRET ANCHORAGES OF BRITTANY 45

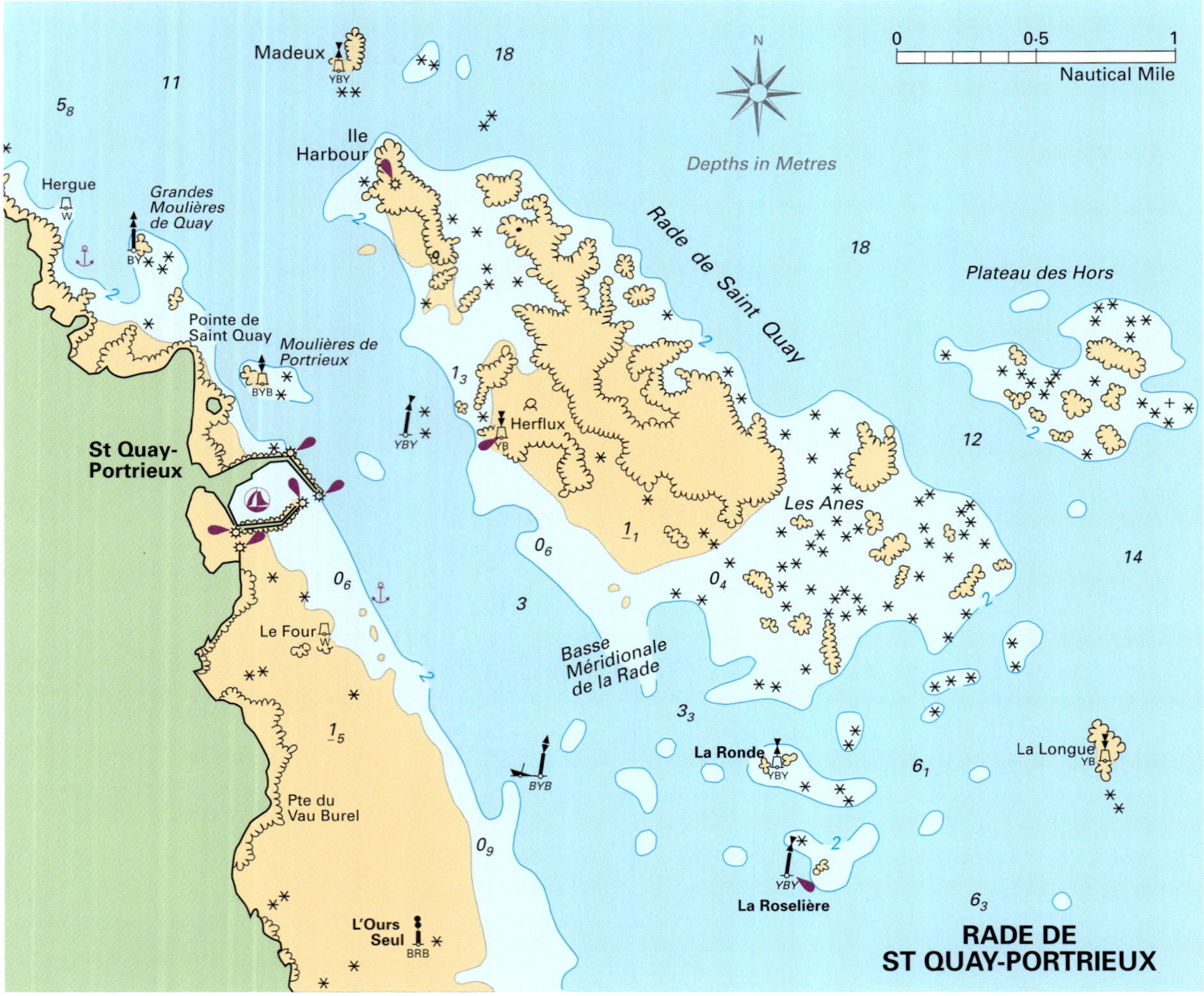

snugly enough half a mile or so ESE of the old Portrieux pierheads, with Le Four white beacon tower bearing southwest about 2–3 cables off. Roches de St Quay offer limited protection in moderate winds from the northeast.

The tidal streams along this rocky stretch of coast flow north-west/southeast through the Rade, only touching two knots during the middle hours of a spring flood. The approaches from either north or south are well marked and straightforward using Admiralty Chart 2029. Coming in from the north, the key marks are Ile Harbour and Madeux west-cardinal beacon tower. From the south, you need to find La Roselière W Cardinal buoy and leave it nicely clear to starboard on the way up to the Rade.

PORT GORET

About two and a half miles northwest up the coast from St Quay-Portrieux, the rather dour anchorage at Port Goret is tucked into the steep coast just west of Pointe du Bec de Vir. You enter from the NNE, keeping the Bec du Vir shore within about 300m to port to be sure of clearing a longish spur of drying rock

CAP DE LA HAGUE TO ANSE DE PAIMPOL

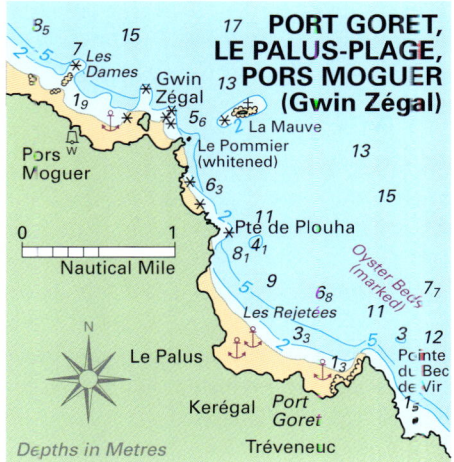

between Pors Moguer and Gwin Zégal, with good holding over sand.

Approach from the northeast, keeping the prominent white daymark on the hill above Pors Moguer bearing 212°. Once the north edge of Gwin Zégal is abeam to port, you can head due south into the anchorage and fetch up as close inshore as the depth allows.

ANSE DE BREHEC

This shallow sandy bay lies just over three miles south along the coast from L'Ost-Pic lighthouse and the Anse de Paimpol. Le Taureau BRB beacon tower is a useful mark, standing on an isolated rock three-quarters of a mile due east of Bréhec. There's also a prominent white tower on the cliffs at Pors Moguer, 1½ miles southeast of Bréhec entrance.

The approach to Bréhec is clean and straightforward, except that numerous small crab-pot floats are usually dotted about in the entrance, south of Beg Min Rouz point. The area is also popular with scuba divers. The inner part of the bay has plenty of local moorings and the

jutting out to starboard. The high dark cliffs and a stark ruined building give this place a spooky atmosphere, but we once rode out a southwesterly gale here in perfect solitary shelter.

LE PALUS-PLAGE

Half a mile WNW of Port Goret, you can anchor in a wider, more congenial sandy bay off the seaside village of Le Palus-Plage, with good shelter in westerlies or southwesterlies. The approach is easy, with the north cluster of houses bearing just south of west true. Be sure to avoid a ledge of rocks called Les Rejetées that jut well out from the shore two cables north of these houses.

PORS MOGUER (GWIN ZEGAL)

About one and a half miles north of Le Palus-Plage, the prominent islet of Gwin Zégal (27m high) is so close to the coast as to seem like a headland. A drying rocky spur juts north from Gwin Zégal for a good 250 metres, and the islet and this spur form the east arm of a snug bay well sheltered from between west and south. On the west side of this bay is the tiny small boat harbour of Pors Moguer and the best anchorage is about midway

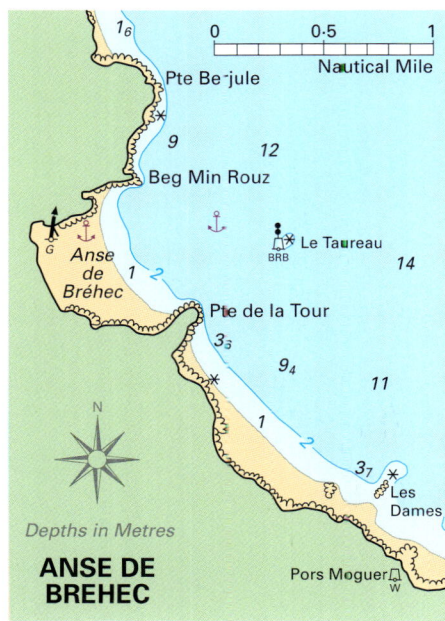

SECRET ANCHORAGES OF BRITTANY   47

small resort of Bréhec lies at its head, with a stub of jetty in the northwest corner.

This agreeable anchorage offers good shelter in winds from NNW through west to south, although a rolling scend often comes in over the shallow bottom with the high tide. You should tuck as close inshore as possible, but most keel boats will need to anchor a quarter of a mile east of Bréhec jetty to stay afloat at low springs.

ANSE DE PAIMPOL

This wide shallow bay is one of the most attractive and atmospheric inlets on the North Brittany coast. Facing east, the Anse de Paimpol is simple enough to approach from that direction in reasonable visibility, between the distinctive L'Ost-Pic lighthouse and Les Charpentiers E Cardinal beacon tower. The streams run strongly NNW/SSE off the entrance, but are generally weak once you are inside L'Ost-Pic, except near Pointe de la Trinité. Most of the bay westward of Ile St Rion dries or is very shallow at LAT, so neap tides provide the best choice of anchorages.

You can anchor a short half-mile SSE of the highest part of Ile St Rion, or a bit nearer the island around neaps, but make sure you are lying outside the withies marking shellfish beds on the wide shoal area south of St Rion. Near neaps, moderate draught boats can anchor close under the low east end of the island, tucking inside the long rocky shoal a little way offshore.

Further into the Anse de Paimpol, you can anchor and stay afloat at most tides in a tongue of water from where Pointe de la Trinité bears north true and tiny Ile Blanche lies a short half-mile to the south. Using your plotter or large-scale chart, Admiralty Chart No 3673, you need to locate the narrow channel that leads towards the head of the bay, picking your way between the Ile St Rion withies to starboard and another extensive area to port. It's not too

L'Ost Pic lighthouse

CAP DE LA HAGUE TO ANSE DE PAIMPOL

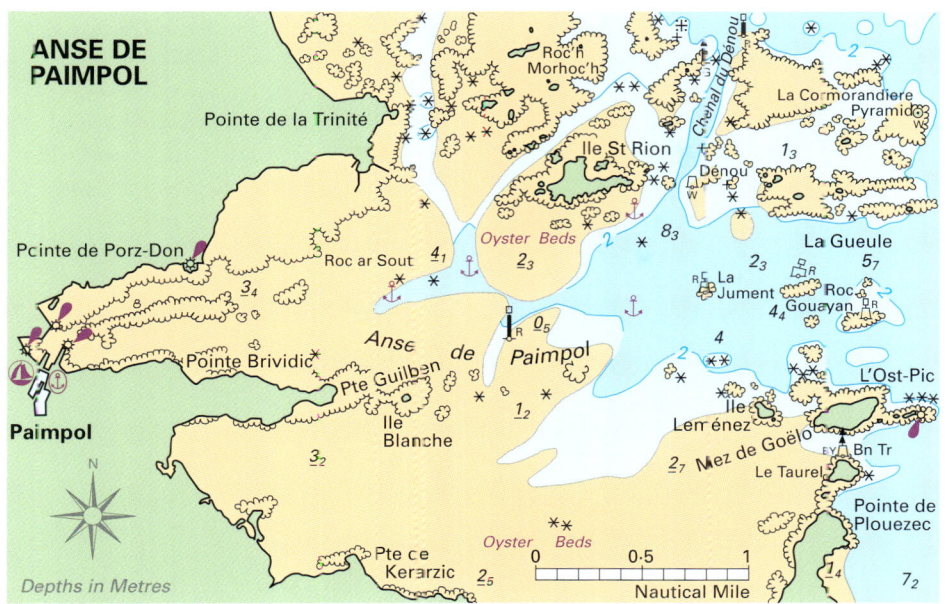

Anchored off the south-east side of Ile St Rion

SECRET ANCHORAGES OF BRITTANY 49

difficult to find this spot near low water if you nudge in very slowly on the echo-sounder.

If you arrive off the Anse de Paimpol near low springs, you can anchor without any delicate pilotage by simply letting go 3-4 cables west of La Jument red beacon tower. As you are coming into the bay from the east, skirt north of Gouayan red beacon tower, Rocher Gueule red can buoy and La Jument, and then keep Rocher Gueule red buoy just open to the north of La Jument. Continue slowly shorewards on this transit, anchoring when you run out of water. If bound for Paimpol, you can start approaching the harbour a couple of hours before local high water.

USEFUL ADMIRALTY CHARTS

No 1114	Approaches to Cherbourg
No 2029	Ile de Bréhat to Cap Fréhel
No 2700	Port St Malo and approaches
No 3653	Guernsey to Alderney and adjacent coast of France
No 3656	Plateau des Minquiers and adjacent coast of France
No 3659	Cap Fréhel to Iles Chausey

USEFUL FRENCH SHOM CHARTS

No 4233	La Rance – De St Malo à l'Écluse du Châtelier
No 7127	Abords de l'Ile de Bréhat, Anse de Paimpol, Entrée du Trieux
No 7129	Du Cap Fréhel à St Briac-sur-Mer
No 7134	Iles Chausey
No 7310	Baie de St Brieuc (partie Est), De Dahouët au Cap Fréhel

Approaching L'Ost Pic and the Anse de Paimpol

THE PAIMPOL COD FLEETS

In the late 18th and 19th centuries, Paimpol grew increasingly prosperous as a cod-fishing base. It's extraordinary to imagine now, but large fleets of the local *goélettes* would *sail* in the depths of winter to the richly stocked cod banks off Iceland and fish away from home for months at a time. February and March were reckoned to be relatively quiet off Iceland and were good months for fishing. The grand departure of the fleet was usually set for 20th February. One of my old Brittany books paints a heroic, rather chilling picture of most of the able-bodied male population of Paimpol setting off on the high seas:

'Under a pale winter sun, a raw February wind filled the sails of the brightly painted fishing boats, decked out with colourful flags to lift the spirits of the twenty or so motley crew, half of whom were not sailors but farm-workers or labourers. For six months at a time, without touching land, these men of Paimpol fished day after day, from dawn to dusk, on a pitiless hostile sea.'

Gutting and cleaning was done at night, so as not to waste valuable fishing time, and the holds were gradually filled with salted cod. Up in those latitudes, especially in winter, well-salted fish would keep for long periods without any problem. The period around the autumn equinox, with its turbulent depressions, was a particularly dangerous time for the Paimpol cod-fishers. Many boats never came back.

It's interesting to reflect that, during the peak years of the Paimpol fishing industry, as many as 80 *goélettes* would have been moored in the basins during the summer, and you could walk from one side of the harbour to the other across the decks. On the far side of Bassin No. 1, the *Musée de la Mer* is situated in the old cod-drying plant on Rue Labenne. This fascinating building houses a permanent exhibition of photographs, artifacts and eye-witness accounts of the long relationship between Paimpol and the sea.

The old tide mill on Ile de Bréhat's west coast

CHAPTER 2

ANSE DE PAIMPOL TO THE PENZÉ RIVER

The 50 miles of coastline between Anse de Paimpol and the Penzé river takes in the whole of the dramatic Côte de Granit Rose and a bit more besides. Although some of the most popular Brittany harbours lie along this stretch, there's still plenty of scope, even in high season, for finding quiet anchorages away from the crowds. This is archetypal North Brittany, starting at the eastern end with the timeless Lézardrieux estuary and the enchanting anchorages around Ile de Bréhat. Both the Lézardrieux and nearby Tréguier estuaries are well peppered with rocky hazards at their mouths, but the rivers themselves mellow quickly as they wind inland between wooded banks to penetrate well into sleepy rural France. Despite the growth of moorings over the years, each has one or two quiet corners where you can drop the hook and connect with the slow natural rhythm of the river ebb and flow.

Cruising westwards from Lézardrieux and Tréguier past Port Blanc, Perros-Guirec, Ploumanac'h, Trégastel and Trébeurden, you can find a string of intriguing coastal anchorages in fine summer weather, some hemmed in by spectacular natural breakwaters of curiously shaped granite. Just south of Trébeurden is the shallow Lannion River, with its sheltered pool at Le Yaudet nicely off the beaten cruising track. This strangely remote river really marks the end of the Côte de Granit Rose. Further west, between Lannion and the wide Morlaix estuary, the cliffs are a more familiar granite grey.

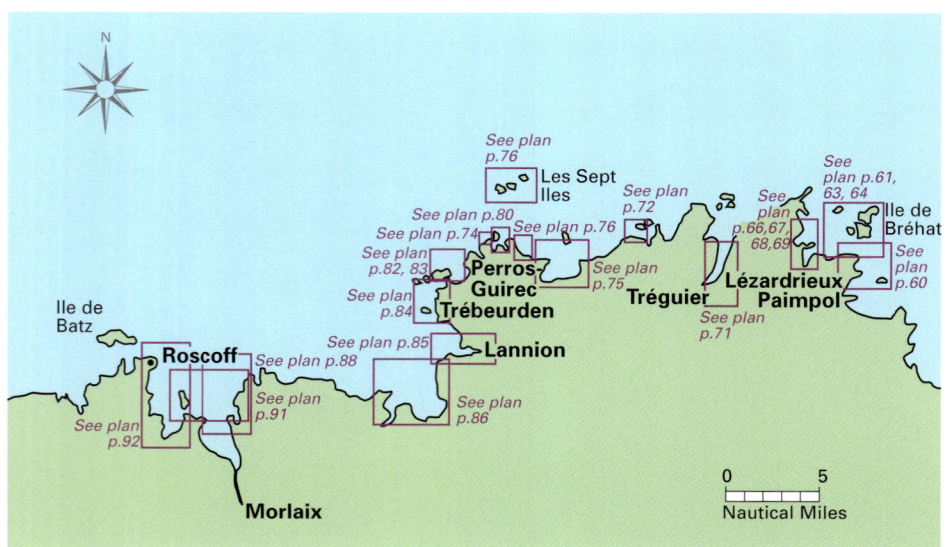

Trieux estuary

Passage-making along this whole length of coast is fairly straightforward, so long as you are carefully rock-conscious and aim to carry sufficient fair tide for each leg. You'll come across the most powerful streams (up to five or six knots at springs) in the approaches to the Lézardrieux estuary between Les Héaux and La Horaine lighthouses, as the tide pours round this jutting corner of North Brittany to fill and empty the Gulf of St Malo. The stream also tends to accelerate in the relatively narrow strait between Les Sept Iles and the mainland near Ploumanac'h, getting up to four knots at springs south of Ile aux Moines during the middle hours of the tide.

Working west under sail from Lézardrieux, it's not usually possible for most boats to carry a full tide right along to the Morlaix estuary, partly because of the longish haul out of Lézardrieux and round Les Héaux to start with. Larger boats can sometimes do it with a cracking fair breeze if they leave a little before high water, but it's more agreeable to make two shorter hops – from Lézardrieux to Perros, Ploumanac'h or Les Sept Iles, say, and thence to Morlaix or Penzé on the next tide.

Swell can be a more significant factor along this coast than in the more sheltered Gulf of St Malo, as you edge further west towards the outer approaches of the English Channel. An Atlantic swell rolling in from due west doesn't tend to carry this far up-Channel close inshore – you usually only start to feel it beyond Ile de Batz. But a partly onshore swell caused by prolonged northwesterlies can find its way into some of the otherwise quite protected anchorages covered in this chapter.

Ile de Bréhat

The atmosphere and character of the natural anchorages between Anse de Paimpol and the Penzé river vary a good deal. On the one hand are the peaceful upper reaches of the North Brittany rivers, where you can fetch up far enough inland to enjoy perfect shelter whatever is happening at sea, yet still stay afloat at any state of tide. Lézardrieux and Tréguier offer the best choices in this respect, and there's nothing quite so soothing, to my mind, as lying at anchor for a few days surrounded by wooded hills, rolling farmland and the placid routines of the country. The subtle sounds and smells of a river often make a welcome contrast to the harsh flavours of the sea, while you know that, at fairly short notice, you can slip downstream to open water and the prospect of new destinations.

One of the small Tréguier estuary buoys just above La Corne lighthouse

La Corne lighthouse

Guerzido anchorage, just west of La Chambre de Bréhat

Les Héaux lighthouse from southward

On the other hand, you have the more weather-dependent coastal anchorages that are perfectly snug in the right conditions but may change mood rather quickly if the wind should shift or freshen. Ile Tomé, Les Sept Iles, Grande Ile, the Trébeurden channel anchorages and Locquirec fall into this category, with Ile Tomé and Sept Iles the easiest and safest to leave after dark if necessary. A night at anchor off Les Sept Iles in the bay between Ile aux Moines and Ile Bono can be an eerie experience. Right next door, the three powerful beams of Ile aux Moines lighthouse sweep the black horizon every 15 seconds. There are no other lights on Sept Iles, so the distant homely glimmer from the mainland at Perros-Guirec somehow enhances the solitude of the anchorage.

Twenty miles southwest of Les Sept Iles, the Morlaix and Penzé estuaries can provide a few secluded anchorages that are only rarely sampled by yachts on passage. Most Brittany visitors know the Pen Lann anchorage at the mouth of the Morlaix River, a safe and convenient spot to wait for the tide before heading upstream to catch the lock into Morlaix basin. You can usually lie comfortably at Pen Lann overnight, reasonably protected from seaward by the headland itself, Ile Louet just to the north, and the advance cordon of reefs out in the estuary. However in calm weather or offshore winds, you can also anchor on the north side of Pen Lann, more or less opposite Carantec in a wide sandy bay. The shallow Anse de Térénez, over on the east side of the Morlaix estuary, can also make an interesting anchorage, especially near neaps. With the wind anywhere in the south, Térénez is better protected than Pen Lann and the small

The inner reaches of Ploumanac'h pool

hamlet ashore has a friendly atmosphere and a couple of good bistros.

Just to the west of the Morlaix estuary, the Penzé is probably one of the least-visited rivers in North Brittany. You need to enter the rather rocky mouth above half-tide, follow the channel carefully and avoid the various patches of oyster bed withies on the way. The reward for this slightly delicate pilotage is peace and quiet, with reliable shelter in almost any weather.

L'ANSE DE LAUNAY

This attractive east-facing bay lies between Pointe de la Trinité and Pointe de l'Arcouest, a little way north of the Anse de Paimpol and a mile south of Ile de Bréhat. You approach Launay by the Chenal de la Trinité, either from the south via Anse de Paimpol or from the north via Bréhat Roads.

Coming from the south, simply follow the charted or pilot book directions from Glividy BRB beacon through Chenal de la Trinité, but make for a position about a cable southeast of Pointe de l'Arcouest once Les Fillettes S Cardinal beacon is within a quarter of a mile bearing north by west. Anchor clear of the local moorings and crab-pot buoys, but make sure by sounding that you are swinging in the tongue of water that doesn't dry at low tide. Coming from the north, enter

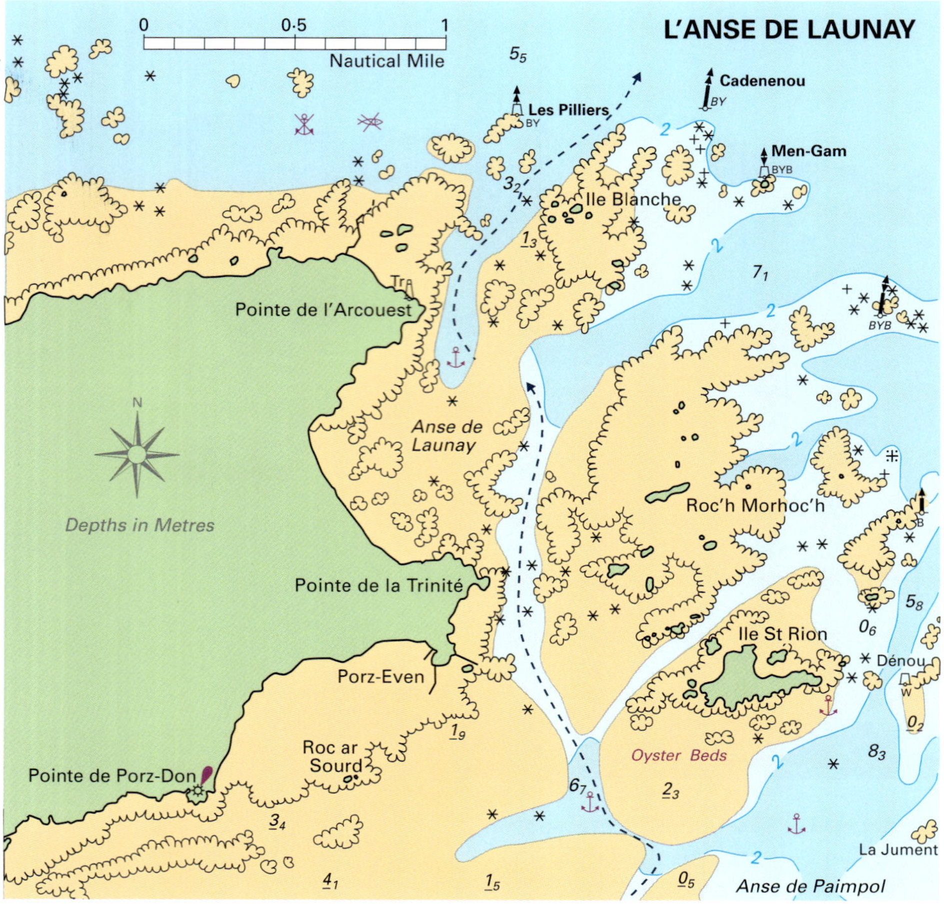

L'ANSE DE LAUNAY

Chenal de la Trinité 1½ cables east of Les Piliers N Cardinal beacon tower and follow the charted directions for less than a mile as far as Pointe de l'Arcouest and the Launay anchorage. Admiralty Chart 3673 shows the way clearly.

L'Anse de Launay is best between neaps and springs in gentle westerly weather, although near low water the off-lying rocks will provide reasonable shelter from between south and southeast. Only stay overnight here if you are sure of the forecast, because it's difficult to leave in either direction after dark.

SOUTHEAST BREHAT

Tucked between Ile Logodec and Bréhat's south-east corner, La Chambre was once a popular anchorage but is now full of local moorings and anchoring is no longer allowed. In quiet or gentle north-westerly weather you may still lie just outside La Chambre's entrance beacons. Between springs and neaps you can anchor further west off the wide shallow bay between La Chambre and Men Alan S cardinal beacon, outside a string of yellow buoys protecting the swimming beach. This is a good lunchtime spot, especially as the tide falls away to expose more enclosing rocks. It is also fine overnight in calm weather.

ANSE DE PAIMPOL TO THE PENZÉ RIVER

LA CHAMBRE AND LA CORDERIE

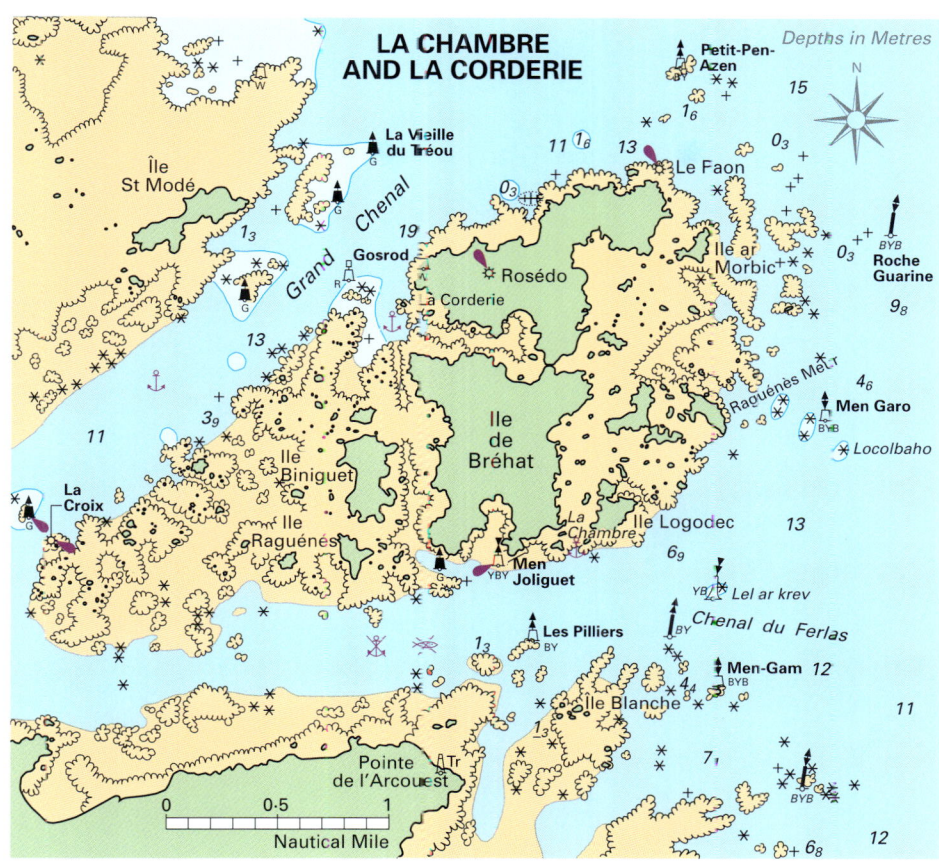

SOUTH EAST BRÉHAT ANCHORAGES

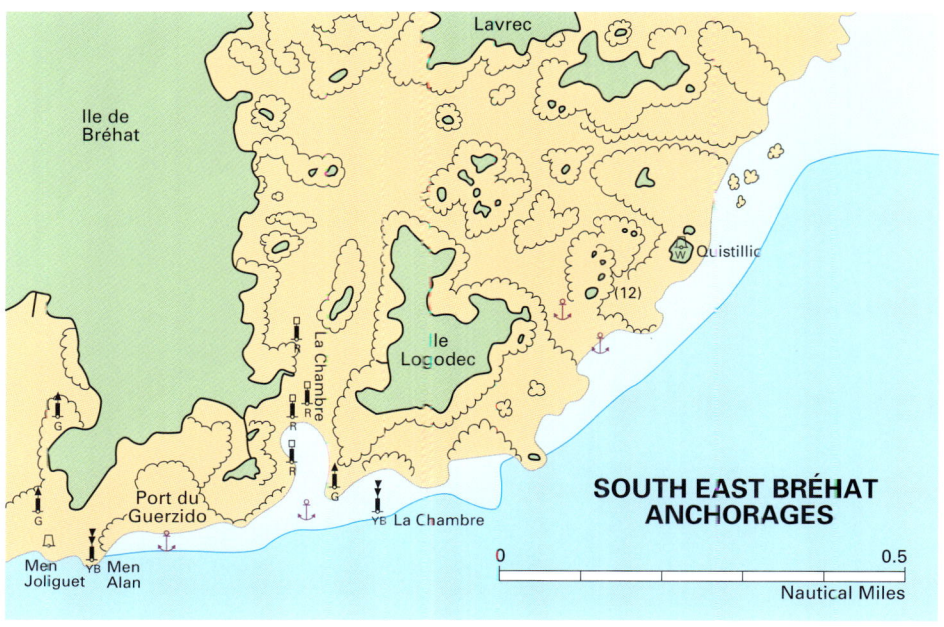

SECRET ANCHORAGES OF BRITTANY 61

COQUILLES ST JACQUES

About eight miles west of Cap Fréhel, the small but bustling fishing harbour at Erquy specialises in gathering local scallops – *Coquilles St Jacques* – from the sandy waters of St Brieuc Bay. The scallops are literally raked from the seabed by fishing boats towing special paraphernalia, and it's common to meet them at work if you are on passage across the bay between Cap Fréhel and Paimpol or Ile de Bréhat. Sometimes you'll see these boats suddenly lurch precariously as their rakes catch on a ledge of rock. The delicate succulent flesh of a scallop is a bit like a nugget of prime monkfish but softer in texture and with a more fishy flavour. At its best, the attached tongue of coral can taste and feel like a parcel of very fine smooth herring roe. Erquy is definitely the best place in which to sample these tasty bivalves. The popular *Coquilles St Jacques gratinées* is no less delicious for being popular, and indeed I much prefer it to some of the trendier scallop recipes. The combination of fresh seafood, cream and finely grated cheese sizzled fast under a hot grill is truly enjoyable, whereas some of the more zany marinades for scallops can leave you wishing you'd ordered something else.

Erquy chefs can do anything with *Coquilles St Jacque*s. In quiet weather near neaps, with high water around lunchtime, you can anchor and stay afloat inside the breakwater in the middle of the harbour, then potter ashore in the dinghy to sample a menu. But don't linger too long over your dessert once the tide starts slipping away. Even a neapy low water uncovers plenty of sand around Erquy harbour.

ANSE DE PAIMPOL TO THE PENZÉ RIVER

ILE LOGODEC

A short half-mile ENE of La Chambre, a shallow drying inlet cuts between the east tip of Ile Logodec and a humped islet 12m high. Between springs and neaps, with the tides taking off, you can nudge carefully towards this gap and anchor as far in as your low water depth allows. Beyond Logodec, a sandy lagoon dries out at springs and in calm weather local boats with bilge or lifting keels sit on the bottom here. So if you approach near high tide, remember that any boats anchored ahead of you may be shoal-draught and expecting to dry.

The ideal time to arrive is near low water when the depths are obvious and you can see your way in between the rocky fringes. When approaching Ile Logodec, keep clear of a line of yellow mussel bed floats usually lying east of La Chambre S cardinal beacon.

LE KERPONT

The narrow Kerpont passage leads close past the west side of Ile de Bréhat and east of the much smaller Ile Biniguet. Near high water, this picturesque channel is a useful short cut between the Rade de Bréhat and La Corderie, but there's also a little-used anchorage just inside the southern entrance to Le Kerpont, opposite the Bréhat lifeboat slip. Because the tide runs strongly through Le Kerpont, it's best to fetch up in this anchorage at reasonably slack water, say within one and a half hours of high or low. Neaps are easier than springs, because the streams are weaker.

You enter the south end of Le Kerpont from just opposite Port Clos, Bréhat's main harbour, and it's best to start from a position less than 100m due south of Men Joliguet S Cardinal beacon, the most prominent mark off Port Clos. From here it's easy to head west into Le Kerpont, leaving La

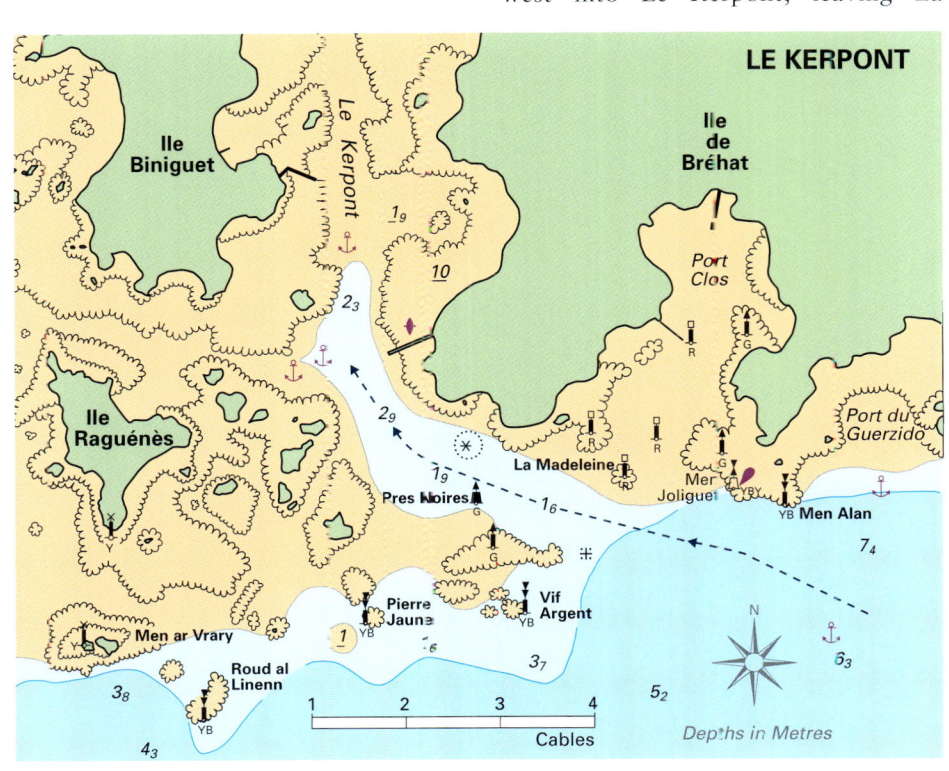

SECRET ANCHORAGES OF BRITTANY 63

Rompa beacon in Chenal du Ferlas

Madeleine red spar beacon about 70 metres to starboard and Pres Noires green beacon a similar distance to port. If you are coming in near low water, be sure to stay well north of an isolated rock awash at datum that lurks 150m SSW of La Madeleine. Off Pres Noires green beacon you start curving *north* round the southwest tip of Bréhat and the lifeboat slip will come into view.

I usually anchor off the west side of the channel about a cable due west of the lifeboat slip, where a slight inlet of deeper water cuts in between two above-water rocks.

LA CHÈVRE

This is a pleasant lunchtime spot on a quiet summer day, best about midway between neaps and springs with low water in the middle of the afternoon. Ideally, aim to approach not too long after high water and get snugly anchored while there's plenty of depth under the keel. You'll then have time to check your anchored position by GPS and cross-bearings before the tide runs away too far.

You approach La Chèvre from Le Ferlas channel, starting from a position a cable due east of Receveur Bihan S Cardinal beacon. From here you can easily identify the islet of La Chèvre (23m high) about four cables to the

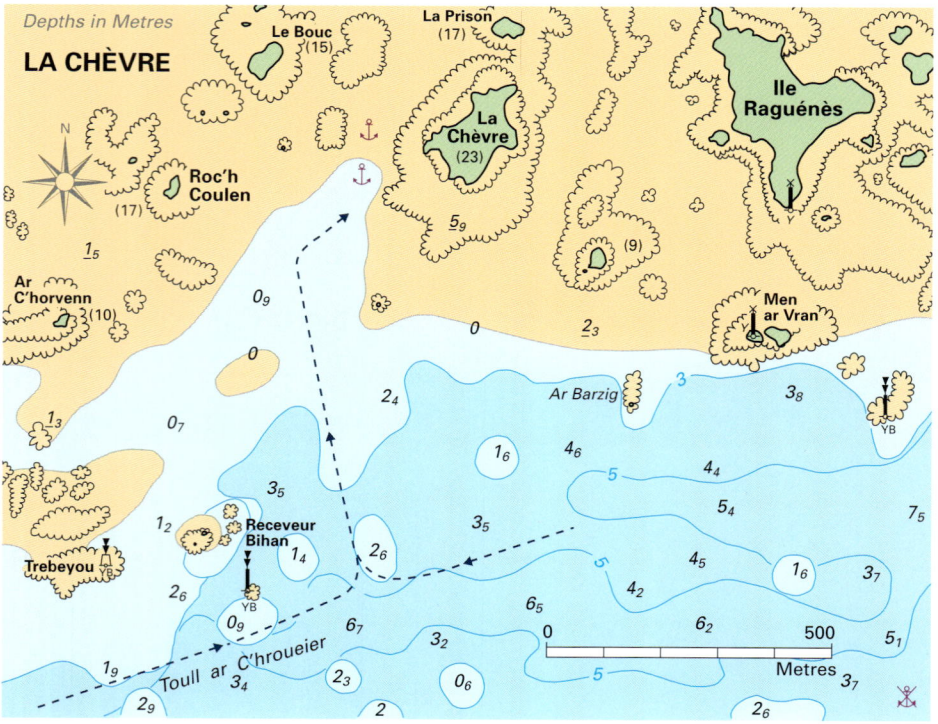

NNW, and the much smaller rock called Le Bouc (15m high) lying 300m WNW of La Chèvre. Bring the highest part of Le Bouc to bear 169° and head straight for Le Bouc on this bearing until the highest part of La Chèvre bears 055° or a shade less. Then turn to starboard towards La Chèvre, steering with the west tip of the islet fine on the starboard bow. Anchor a short cable WSW of this west tip, over the 0.3m datum sounding shown on Admiralty Chart 3673. Between springs and neaps, most boats will stay afloat here at low water and be well sheltered by the surrounding reefs and shoals.

LA CORDERIE

This fascinating long inlet on the west side of Bréhat feels wild and snug at the same time. Most of La Corderie dries at springs, but the neap anchorage near the

La Corderie, Ile de Bréhat

La Croix lighthouse

head south for a quarter of a mile, between Men-Robin green spar beacon and the prominent Kervarec Rock. Once past Kervarec, turn ESE to enter La Corderie between two red spars and a green spar.

At springs, keel boats must fetch up just inside these outer beacons, but at neaps you can lie further in, towards or just beyond the next green. Bilge-keelers or boats with legs can take the ground on firm sand right in the northeast corner of La Corderie, one of the most secluded spots on the North Brittany coast.

ILE ST MODÉ

This low, rather mysterious island lies off the west side of the outer Trieux estuary, opposite the north end of Bréhat. The island itself is private, but in gentle westerly weather you can lie snugly a cable off its east shore, nicely out of the main estuary tide and especially sheltered near low water.

entrance is sheltered from southwest through south to northeast. Moderate westerlies present no real problem, since there's only a limited fetch across the Trieux estuary, but northwesterlies and northerlies can send in a swell above half-tide. Approach La Corderie from the Grand Chenal by first making a position 200-250 metres due west of Rosédo white pyramid daymark. Then

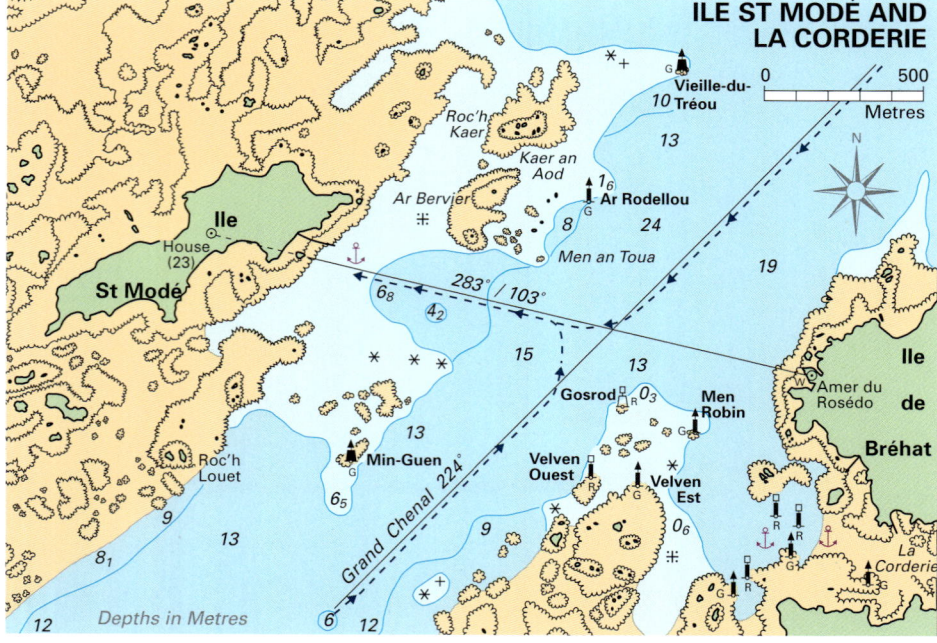

ANSE DE PAIMPOL TO THE PENZÉ RIVER

Approach Ile St Modé from the Grand Chenal leading line (La Croix lighthouse on with Bodic tower bearing 224°), turning off WNW when you are on a straight line between the old watch tower on the island and Amer du Rosédo white pyramid on the northwest tip of Bréhat. Then head for Ile St Modé, keeping the watch tower bearing 283° ahead and Rosédo bearing 103° astern. The watch tower will be more or less in transit with the left-hand edge of a sea wall on the northeast shore of St Modé.

Anchor off the old St Modé landing jetty over the 1.3m datum sounding shown on Admiralty Chart 3673. I have stayed here overnight in quiet weather and this spot is easy to leave in the dark if necessary. Simply come out with Rosédo light bearing 100° until you join the transit of the Grand Chenal leading lights. Then it's a simple matter to scuttle upriver towards Lézardrieux.

RADE DE POMMELIN

This apparently rather open roadstead branches off the west side of the lower Trieux estuary about four cables northwest of La Croix lighthouse. In this unlikely spot you can find reasonable shelter and perfect seclusion in moderate westerly weather, with good holding in sand and mud. Coming up the Grand Chenal, simply bear to starboard about half a mile before reaching La Croix, steering to pass a cable north of the N Cardinal spar beacon that guards the northwest corner of Plateau de Moguedhier. Then carry on a little further WSW to anchor somewhere between Guazec Guen red spar beacon and Hole Vras green spar beacon. At neaps you can edge a little further WSW along the narrow tongue of deeper water shown on Admiralty Chart 3673.

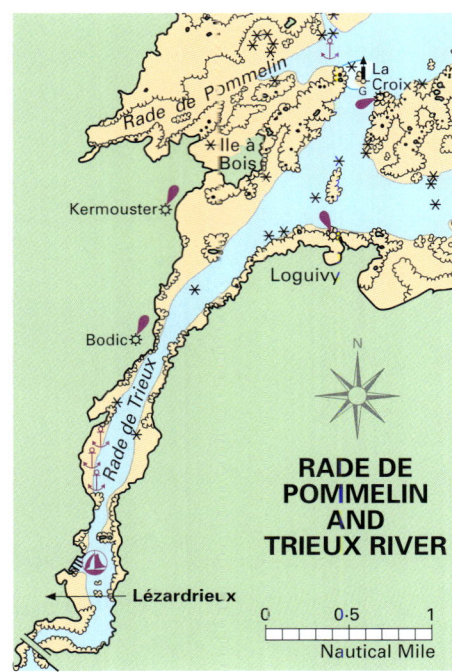

TRIEUX RIVER

Many yachts visit the Lézardrieux marinas and visitors' buoys each season, but the river itself still offers one or two quiet, sheltered anchorages in idyllic surroundings, despite the growth in the numbers of moorings.

Bodic We sometimes anchor in the lower river close off the west shore just above Vieille de Bodic green beacon tower. At dead neaps you can tuck in

The Trieux estuary above Lézardrieux

SECRET ANCHORAGES OF BRITTANY 67

ANSE DE PAIMPOL TO THE PENZÉ RIVER

The road bridge just above Lézardrieux

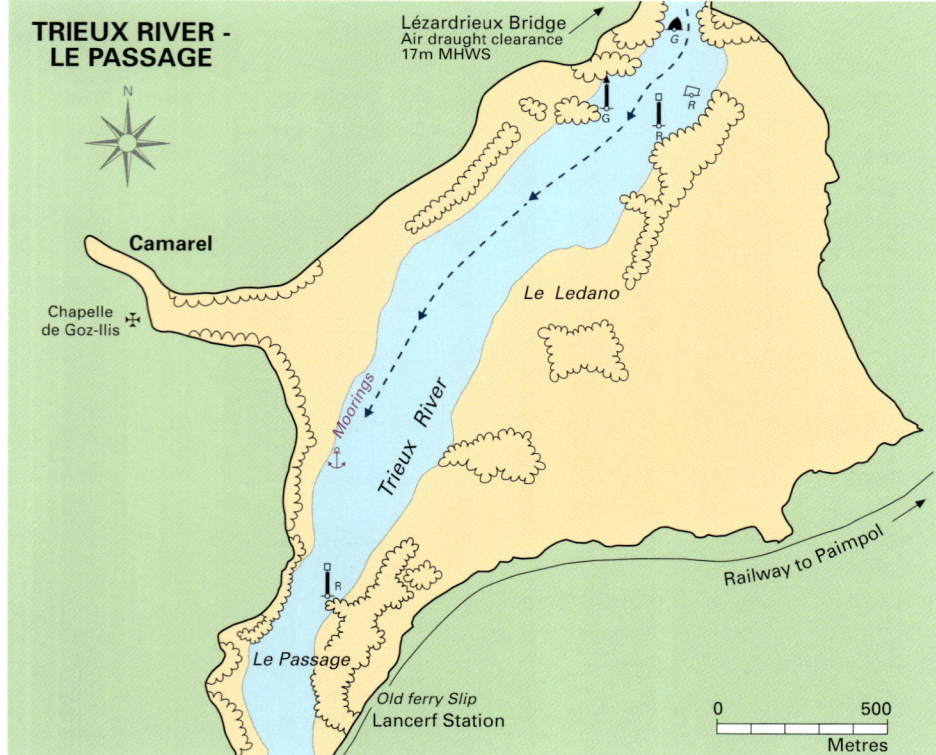

quite close here and stay afloat, with Vieille de Bodic bearing ENE about a cable off.

Perdrix At neaps, moderate-draught boats can usually find room to swing half a mile below the Lézardrieux marinas off the west bank, 1–2 cables downstream from Perdrix light and just inside the local moorings. At springs you can lie outside the moorings, although the tide runs quite hard here and you'll be rather out in the fairway.

Le Passage Leaving Lézardrieux a couple of hours before HW, you can comfortably follow the river round the corner to Lézardrieux bridge (17m clearance). Above the bridge the channel dog-legs past a pair of beacon towers and then the river widens considerably for almost a mile. The fairway isn't marked across this expanse, but you should set off at about 210°, keeping

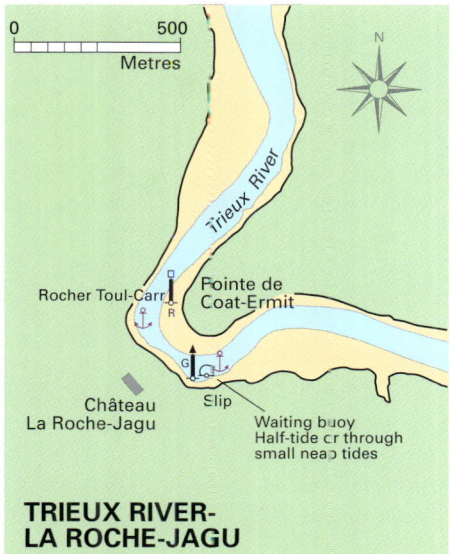

The upper Trieux River at Pontrieux basin

ST YVES OF TRÉGUIER

'Monsieur Saint Yves', Tréguier's most notable Breton saint, was a popular working lawyer during the 13th century. Yves Hélory de Kermartin was born in 1253 to an aristocratic family living just outside Tréguier. Having studied law at Orléans and Paris, he returned to Brittany and took Holy Orders. Yves became an ecclesiastical judge renowned for his fairness, and he developed a reputation as a medieval Solomon. He was also chaplain to the Bishop of Rennes. Spurning honours and preferment and always refusing bribes, Yves championed the poor, widows, orphans and the disadvantaged. Apparently he was also very intolerant of any litigants who didn't tell him the truth. One of the earliest architects of free legal aid, Yves was altogether an all-round honest advocate – not a common species at that time.

After his earthly death in 1303, it was less than fifty years until Yves was canonised and he became the patron saint of Brittany as well as of lawyers and judges. The Pardon of St Yves takes place each year at Tréguier on the Sunday nearest 19th May and is attended by members of the legal profession from all over Europe. This annual pilgrimage is also known as the Pardon of the Poor, a reference to St Yves rigorous spirit of justice rather than to any patronly professional support from the prosperous advocates who gather to take part in the mass at Tréguier cathedral. Afterwards, all these fully gowned legal eagles join a colourful procession of dignitaries through the town and out along the lanes to the neighbouring village of Minihy-Tréguier.

This spectacular *pardon* is one of the largest in Brittany, with thousands of people flocking into the Tréguier area for the long weekend. Even if you are not interested in *pardons*, this date is worth noting if you are cruising hereabouts, because many of the local restaurants will be fully booked!

broadly to the west side. Before the river narrows again at Le Passage, there's a secluded anchorage off the marshy west bank near some small boat moorings. Most boats can stay afloat here up to halfway between neaps and springs.

La Roche-Jagu In the upper reaches of the Trieux River, a couple of miles above Le Passage, you can stay afloat around neaps in a tight bend just beneath the imposing Château de la Roche-Jagu. Tuck close in under the west bank for the deepest water. To reach this memorable anchorage, it's best to leave Lézardrieux a couple of hours before HW.

TRÉGUIER RIVER

There are various good anchorages in the lower reaches of the river if you don't want to go all the way up to the marina. In westerlies, you can anchor 3–4 cables southeast of Ile Loaven, on the west side of the channel between the first two red buoys. Further upstream is a sheltered spot off the west bank between No. 1 green conical buoy and Roche Don green spar beacon.

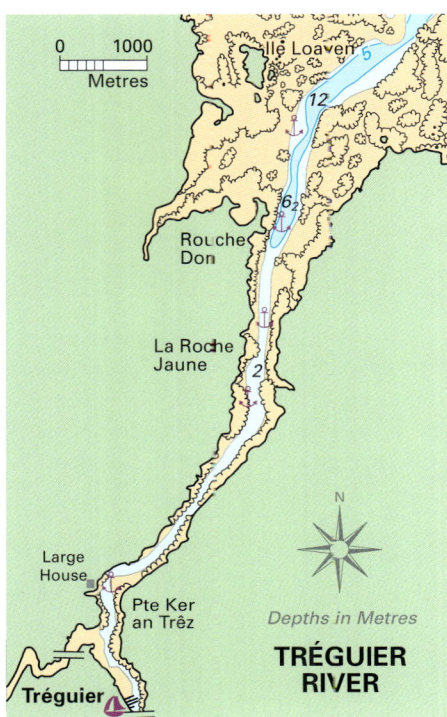

Tréguier River

ANSE DE PAIMPOL TO THE PENZÉ RIVER

Local sculling in the Tréguier River

La Roche Jaune is a pleasant village on the west side of the Tréguier River, about one and a half miles above the first red buoy. You can still anchor slightly downstream of Roche Jaune, clear of the local moorings, where there's good shelter in all but fresh northeasterly winds. There's another possible spot about half a mile upstream from Roche Jaune, between No. 5 green buoy and No. 6 red. You must stay near the middle of the channel here because the sides dry out to steep banks of mud.

A useful and attractive anchorage for deep-draught yachts is in the bight of river just below No. 10 red buoy, opposite Pointe Ker an Trêz. Fetch up close under the wooded northwest shore, with the south side of the large château-like house bearing about 250°.

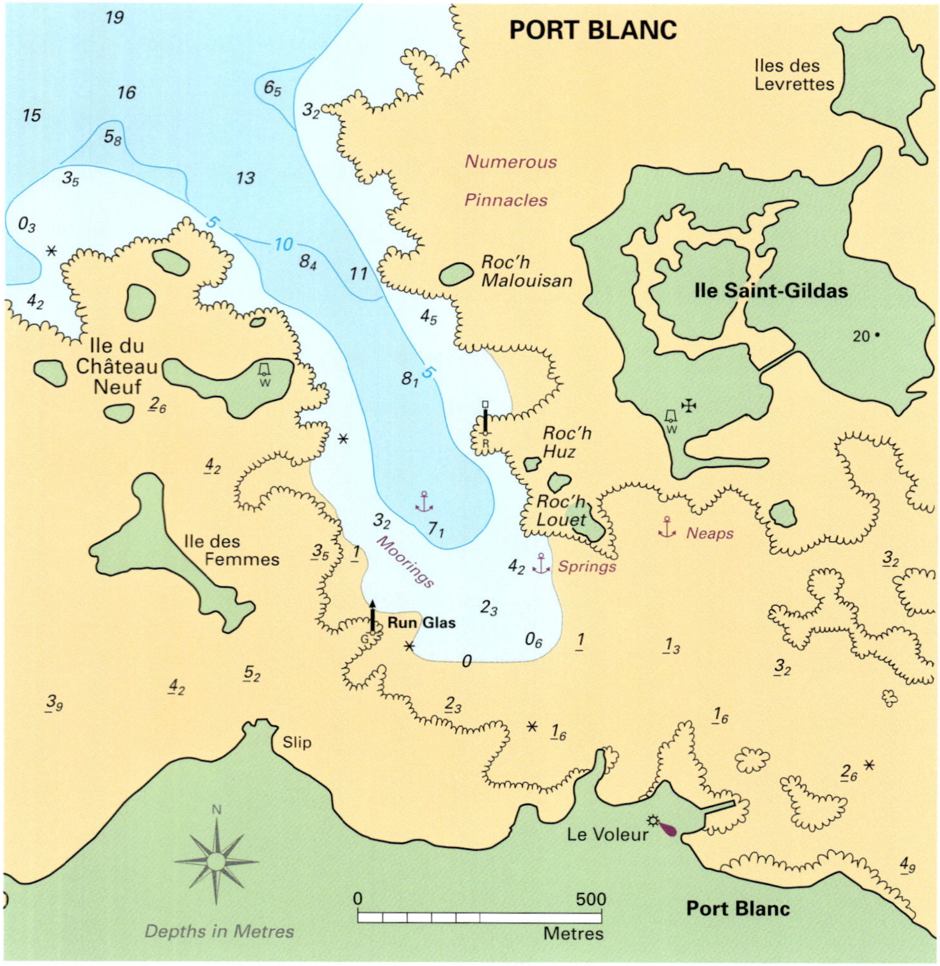

PORT BLANC

This small fishing village and resort huddles behind a cordon of rocks between Tréguier entrance and Perros-Guirec, providing an interesting area to anchor in quiet summer weather or in winds with any south in them. Port Blanc really comes under the 'harbour' category and there are directions in any of the pilot books, but I have shown here a good spot where boats of moderate draught can lie snugly at dead neaps, perhaps just touching bottom at low water. By edging in close to the south tip of Ile St Gildas, you can gain some extra shelter from swell under the islet of Louet, especially as the tide runs away.

ANSE DE TRESTEL

About halfway between Port Blanc and Perros-Guirec, this splendid sandy bay offers an intriguing daytime anchorage in quiet summer weather, idyllic for swimming and a long lazy lunch. The beach is easily identified by the prominent building at its east end. It's best to approach near high water, starting from a position on the Anse de

Port Blanc

The rocky channels at Trégastel

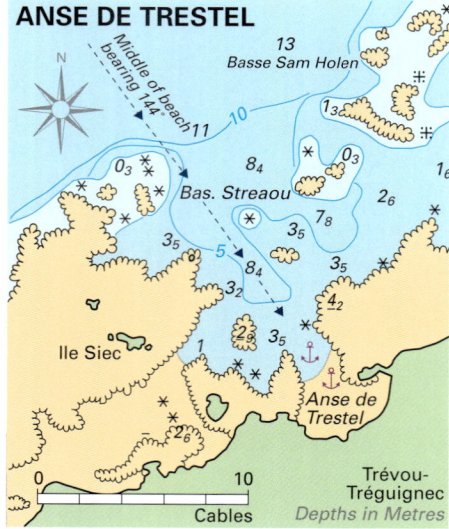

Perros Passe de l'Est leading line with the north edge of Ile Tomé bearing due west true.

From this clear water position you can head directly for Anse de Trestel, keeping the middle of the sandy beach bearing 144° exactly. This line leads safely in through various rocky shoals, leaving a prominent above-water rock about 250m to starboard. The low plateau of Ile Siec will be left a short half mile away to starboard. Keep heading for the middle of the beach and anchor at first about 250m WNW from the tip of the short rocky promontory on the east side of the bay. As the tide falls away, you can take careful soundings and perhaps move further in towards the beach near low water, when the anchorage becomes increasingly sheltered by Ile Siec and its surrounding reefs and shoals.

74   SECRET ANCHORAGES OF BRITTANY

ILE TOMÉ

This narrow but quite high island, shaped rather like a fish on the chart, lies a mile offshore in the approaches to the shallow Anse de Perros. There are various positions, depending on the state of tide, where you can anchor off its east side, but I usually make for a spot a couple of cables off the southeast shore, where the holding is more predictable than further north. Be sure, in any case, not to anchor over the rocky ledge known as Platier de Tomé (0.6m at LAT) that juts well out halfway along the east coast.

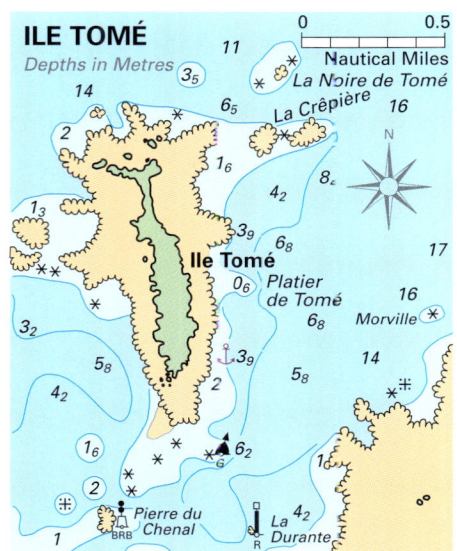

ANSE DE PERROS

In quiet summer weather, you can anchor overnight a mile or so seaward of Perros harbour entrance, 3–4 cables southeast of Pointe du Château (marked Castell Perros or Kastel Perros on some charts) and a quarter of a mile NW of Roc'h Hu de Perros red beacon tower. This spot is fairly well protected from the west through south to southeast,

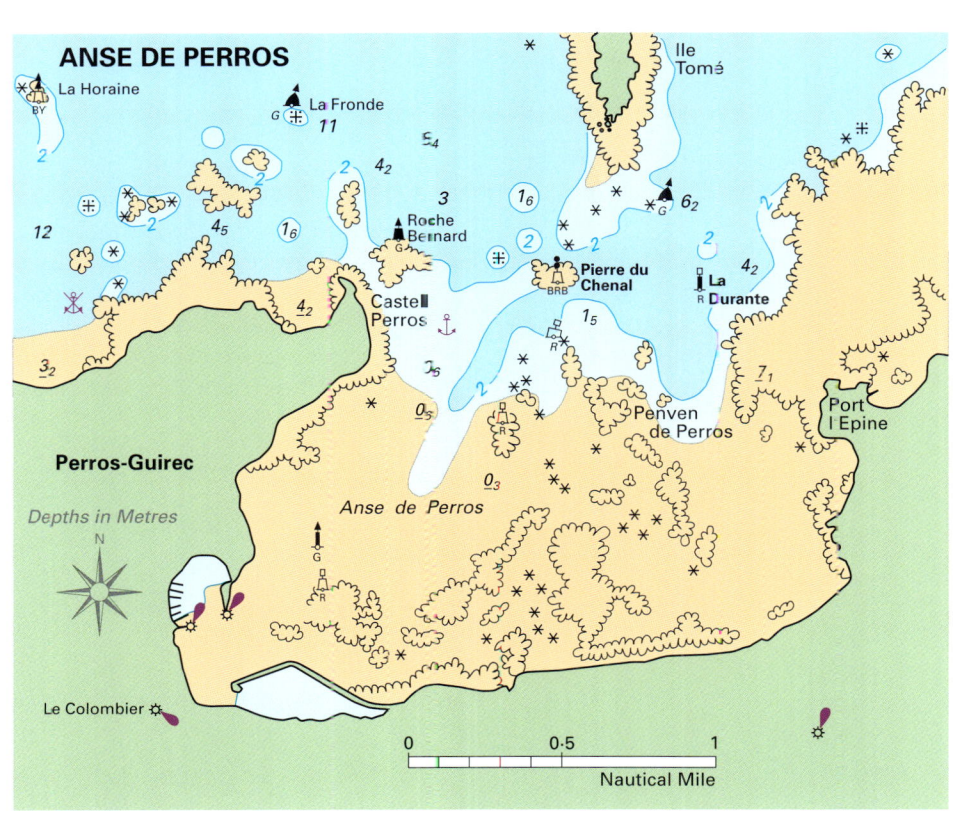

ANSE DE PAIMPOL TO THE PENZÉ RIVER

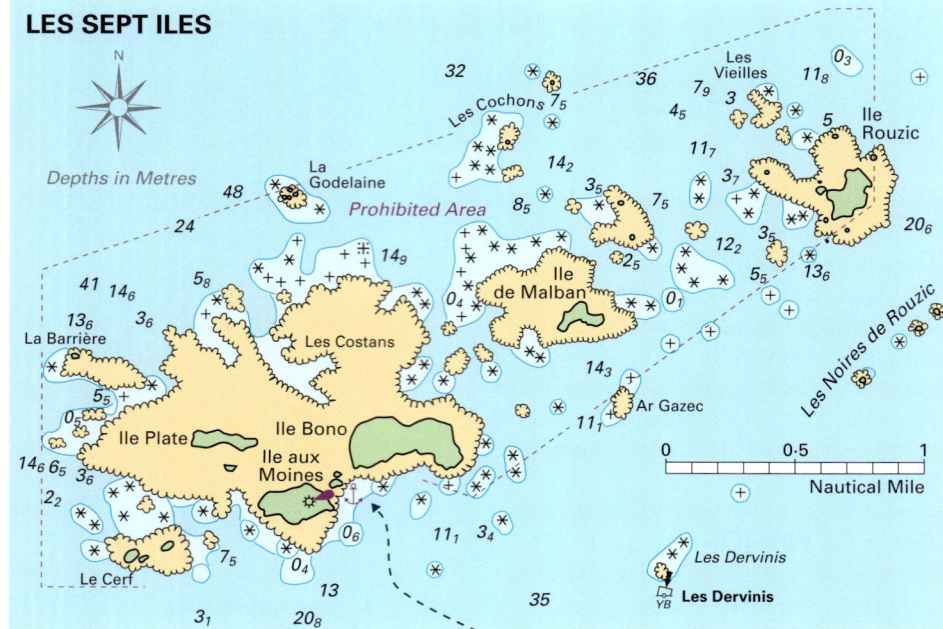

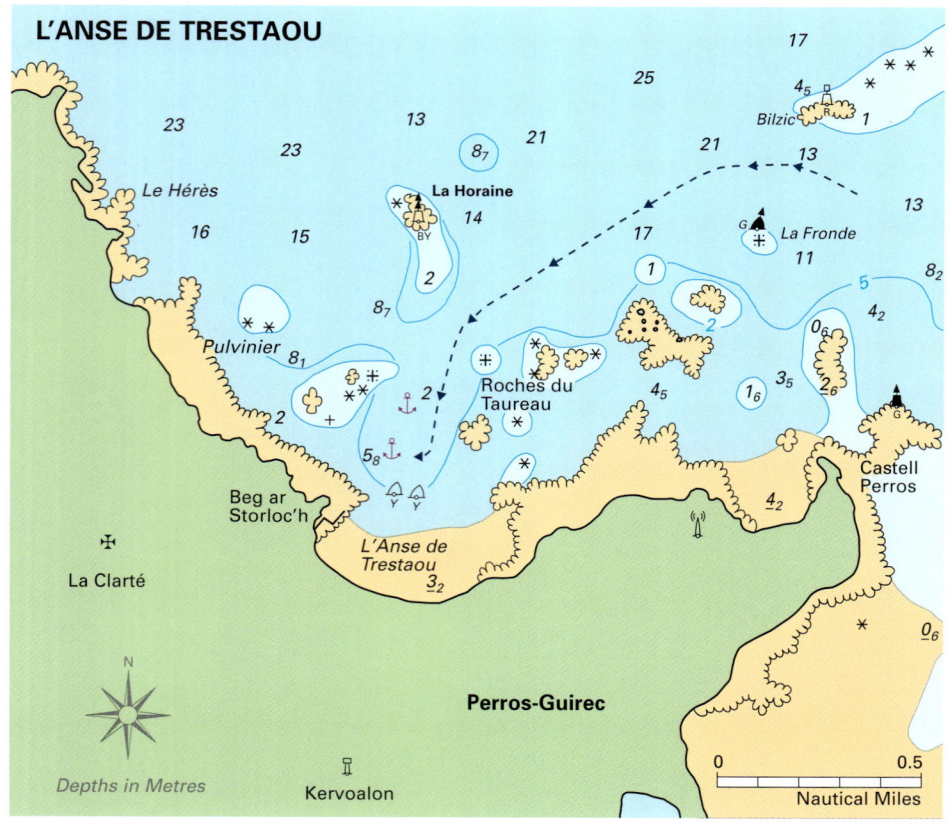

although there's usually a slight scend rolling in for a couple of hours either side of HW. The anchorage practically dries at LAT, but most boats can just about stay afloat here at MLWS.

LES SEPT ILES

There's an attractive fair weather anchorage in the bay formed by Ile aux Moines and Ile Bono. Above half-tide the various shoals in the southern approaches are safely covered and you can simply approach from the SSE, steering towards the western half of Ile Bono. Nearer to low water, though, you need to avoid various rocky patches with more marginal depth over them, including an isolated rock awash lurking eight cables west of Les Dervinis reef and about half a mile southeast of Ile aux Moines lighthouse. In this case, approach on a line with the lighthouse bearing 320° and then alter to head for the west half of Ile Bono when Les Dervinis S Cardinal buoy bears due east true.

Anchor opposite the west end of Ile aux Moines with the lighthouse bearing between 255°–260° and each island shore about 250m off. An overnight stay is not recommended unless the weather is quiet and settled, although it's not too difficult to leave here in the dark if necessary and scuttle over to the well-lit Anse de Perros.

L'ANSE DE TRESTAOU

This wide sweep of bay west of Perros-Guirec is fringed by a magnificent white beach and several large buildings, including a prominent casino. The bay is well littered with rocky dangers for three-quarters of a mile offshore, but in quiet summer weather or offshore winds a couple of simple transits can lead you in safely to a lunchtime anchorage off the west end of the beach – the popular Plage de Trestaou.

The easiest approach is if you are coming out from Perros through the Passe de l'Ouest. Once you have left La Fronde green buoy to port, head pretty much due west true towards La Horaine N Cardinal beacon tower until, over on your starboard quarter, the north edge of Ile Tomé bears 058° in line with Bilzic

Ploumanac'h approach channel

SEABIRDS AROUND LES SEPT ILES

Often when we cross to North Brittany from Dartmouth, Les Sept Iles are the first land to emerge with the dawn after an overnight passage. In good visibility and low early morning light, their distinctive humps stand out sharp and black against the Brittany mainland, tailing away eastward to a string of jagged reefs. To the west you can usually see the stubby tower on Plateau des Triagoz, out on its own where the coast falls away towards Roscoff and the Bay of Morlaix. The largest of Les Sept Iles, Ile aux Moines, lifts steadily out of the sea, easily identified by its prominent lighthouse.

Lying just over two miles north of Ploumanac'h, this unique cluster of small islands and reefs is home to a prodigious seabird population. Because the whole area is preserved as a bird sanctuary and nature reserve, landing is only allowed on Ile aux Moines, named after a settlement of monks that lived here during the Middle Ages. Much later, a garrison was stationed on the island after Louis XV ordered a fort to be built there. Today only the lighthouse maintenance staff stay on Moines regularly. The lighthouse was built in the 1830s, destroyed in the Second World War and restored in the late 1940s. During the 19th century thousands of puffins used the islands as a breeding ground, but with the increase in tourism in the late 1800s and early 1900s, hunting became popular and puffin numbers were decimated. In 1912 the Ligue pour la Protection des Oiseaux (LPO) obtained a ban on hunting and also on people landing on Ile Rouzic, the most easterly of the islands.

After Les Sept Iles became a reserve in 1912, puffin numbers recovered and from just 400 pairs in 1912 the population rose to over 7,000 pairs by 1950. Other seabirds started colonising the islands and in 1939 the first gannets were recorded breeding on Rouzic. In 1976 Les Sept Iles were designated a full nature reserve by the French Government. The islands are now home to breeding pairs of storm petrel, shags, Manx shearwater, fulmar, oystercatchers, puffin, razorbill, guillemot, herring gull, greater and lesser black-backed gull. Gannets nest on Ile Rouzic, the most easterly of the islands. Other birds seen around Les Sept Iles are kittiwake, common terns and a pair of ravens. Grey seals also breed here and the young pups are easier to see than their parents as they bask on the rocks because their coats are a paler colour and don't merge in with the background so easily. Although you can anchor in the bight between Ile aux Moines and Ile Bono, perhaps the best way to experience this wonderful nature reserve is on a *vedette* trip from Perros-Guirec or Ploumanac'h. The skippers know every rock and crevice and the guides provide a commentary in several languages pointing out the different species and the history of the islands. April or May is usually the time you will see the greatest variety of birds, with Ile Rouzic looking white with the thousands of pairs of gannets nesting there.

ANSE DE PAIMPOL TO THE PENZÉ RIVER

The château on Ile Coastaérès, Ploumanac'h

red beacon tower. Now turn southwest to keep this stern transit exactly in line for about half a mile, until Kervoalon water tower bears 014° and is nicely open to the west of the prominent Roches du Taureau. Follow this line towards the beach and anchor outside the cordon of yellow buoys guarding the swimming area.

PLOUMANAC'H

This amazing inlet through a chaos of pink granite is easier than it looks on the chart and is well covered in the pilot books. You should only approach Ploumanac'h by day and in reasonably quiet conditions. Entry is also much simplified if you arrive above half-tide. Make the final approach to the outer green spar beacon from a little east of north. Leave this outer green beacon close to starboard and then follow the port-hand red spar beacons and another green at about 214°, leaving the prominent Château Costaérès to starboard. The outer anchorage area lies in the stretch of channel between the first and fourth red beacons. In season you'll see a few moorings, at which most yachts should just stay afloat around neaps. At springs you can anchor and stay afloat between the first two red beacons.

TREGASTEL (COZ PORS)

About 1½ miles west of Ploumanac'h entrance, a fairly straightforward channel leads south through the coastal reefs between red and green spar beacons to the Coz Pors anchorage off Trégastel. Although the approach looks rather exposed on the chart, Coz Pors is surprisingly well protected by its off-lying rocks and is particularly snug in southerlies or southeasterlies. You can

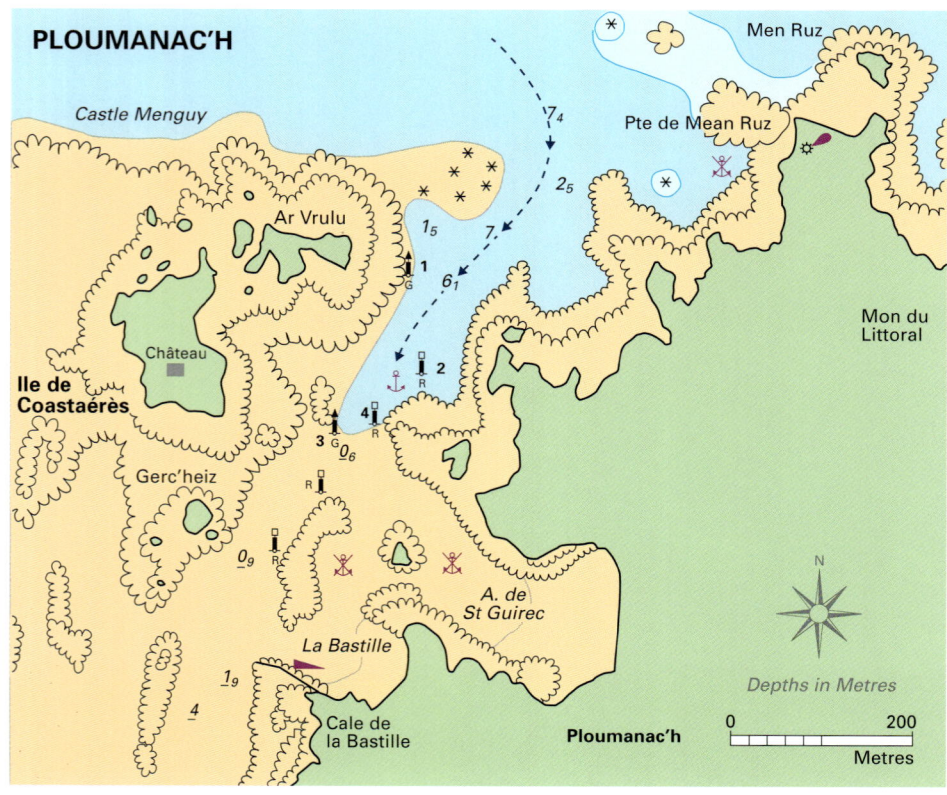

ANSE DE PAIMPOL TO THE PENZÉ RIVER

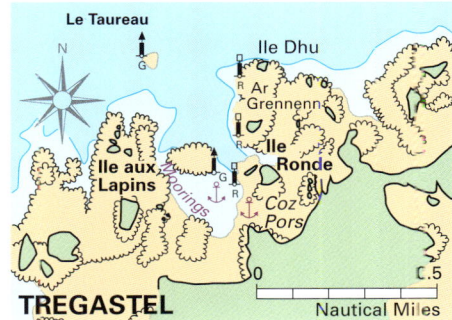

enter or leave at any state of tide, although not at night.

The entrance lies just west of Ile Dhu, an easily identified islet with a red spar beacon on its west side. A quarter of a mile WNW of Ile Dhu, Le Taureau green beacon stands on an isolated rock that dries 4.5m. Approach from seaward steering midway between Ile Dhu red beacon and Le Taureau green. There are two more red beacons south of Ile Dhu and a green beacon close northwest of the third red. It's easiest to follow the red beacons south, entering Coz Porz between the third red and its opposite green. At neaps you can tuck behind Ile Ronde and stay afloat, but near springs you'll have to lie just inside the inner pair of beacons. There are now a good many moorings in Coz Porz and you should always buoy your anchor.

ANSE DE MORVILLE

About midway between Trégastel and Tréburden, this fascinating anchorage off the northeast side of Ile Grande is little visited by cruising yachts because the rocky coast off this low corner of Brittany looks so discouraging. However, in quiet weather or winds from between south and southeast, Anse de Morville is much easier to enter than it looks.

Using Admiralty Chart No. 3669, first reach a position close west of Bar all Gall W Cardinal buoy, preferably near a neap high water when the stream is slack. With plenty of rise of tide, it's then easy enough to make good southeast for a mile towards two above-water rocks – Men Haer (sometimes called Mean Gaez) and Le Corbeau. Men Haer lies half a mile northwest of the northeast tip of Ile Grande and Le Corbeau is a little further inshore.

Keep the prominent white radome (two miles inland behind Ile Grande) bearing 129° just open to the east of Men Haer and Le Corbeau, a simple transit that clears the various rocks on the port hand. As you come within a cable of Men Haer, steer to leave it and Le Corbeau close to starboard and then continue southeast between Ile Grande and Ile Morville to enter the shallow pool opposite the stone landing jetty off Ile Grande. At springs you can stay afloat in the narrows midway between Ile Grande and Ile Morville, but at neaps you can lie a couple of cables further in to the east.

ILE LOSQUET

In settled easterly weather, there are several interesting anchorages to the west and southwest of Ile Grande, best approached near HW neaps for a first visit, when the tide is slack. To reach the Ile Losquet anchorage, first make a position at least a mile west of the island

Plateau des Triagoz

SECRET ANCHORAGES OF BRITTANY

ANSE DE MORVILLE AND ILE LOSQUET

and bring the versatile white radome on the mainland to bear 095° exactly. Close the coast on this bearing, with the highest point of the low-lying Ile Fougère in transit with the radome. This line leads safely between Le Four rock (dries 3.6m) and Ar Bommello (dries 3.9m), and then close north of Morguen N Cardinal beacon. Depending on wind direction, you can either turn north to anchor off the southeast corner of Ile Losquet, or continue on the transit towards Ile Fougère and anchor close off its west shore.

KARREG AR MEG

Above half-tide, you can round Ile Fougère to the south, leave Karreg Jentil S Cardinal beacon close to port and then edge NNE towards Karreg ar Meg E Cardinal beacon tower. At dead neaps, yachts of moderate draught will just stay afloat slightly northwest of a line between Karreg Jentil beacon and Karreg ar Meg tower. This spot is reasonably well sheltered in any gentle summer weather, especially as the tide falls away, but Ile Grande and the mainland give the best protection in easterlies or southeasterlies.

ANSE DE PAIMPOL TO THE PENZÉ RIVER

About a quarter of a mile northeast of Karreg ar Meg is the long stone landing jetty at the southwest tip of Ile Grande. Lifting keel boats or bilge-keelers of modest draught can find a snug and peaceful berth by approaching the jetty near high water, anchoring near the local boats and drying out on the wide area of firm sand off the end of the jetty. There's a small village on the island, about a kilometre along the lane from the landing.

ILE MILLIAU

This attractive island lies close west of Trébeurden marina and acts as a natural breakwater to help shelter the marina entrance. However, even with the marina so close, it's worth remembering that you can still anchor off the north side of Ile Milliau, clear of the local moorings. A slight roll comes in near the top of the tide even in calm weather, but it's easy to enter the marina from here if necessary, or clear out to sea. You can

KARREG AR MEG AND ILE MILLIAU

SECRET ANCHORAGES OF BRITTANY   83

ANSE DE PAIMPOL TO THE PENZÉ RIVER

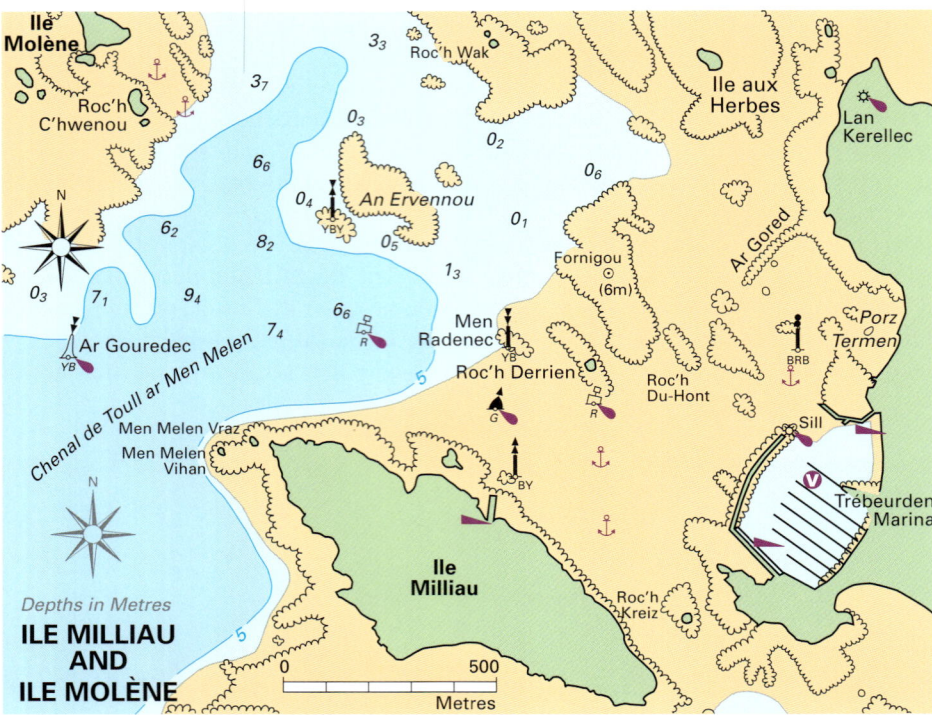

The glorious sandy
beach on Ile Molène

land at the jetty about halfway along the north shore and enjoy a pleasant stroll round the island, with some interesting views east towards Trébeurden and southeast towards the Lannion River.

ILE MOLÈNE

This low rugged-looking island lies a short mile northwest of Trébeurden marina entrance and you can anchor off a splendid beach about 200m off its southeast corner. Approach from the SSE from a position a cable west of An Ervennou W Cardinal beacon, ideally a couple of hours before low water when the various reefs east of Molène are clearly exposed. Between springs and neaps fetch up about 50m east of Roc'h C'hwenou, but you can edge closer inshore towards Molène at neaps. In any case, work out your low water depth carefully and don't swing on too wide a scope.

LANNION RIVER

The entrance lies two miles south of Trébeurden and dries at datum. Approach from due west an hour or two before HW, leaving Le Taureau rock half

Green beacon in the Lannion estuary

a mile to the north, Locquémeau green buoy half a mile to the south and Kinierbel green buoy about two cables to the south. Having passed the Kinierbel buoy, carry on due east towards Beg Léguer lighthouse for another half mile until the inner of the two green beacon towers off the south shore of the estuary is just open to the east of the outer. Turn to the southeast to follow this transit (bearing about 130°) into the river, leaving the two beacon towers each 100 metres to starboard and following the curve of the south shore gradually round to port. This wide sweep avoids the sandbank that extends well out from the north shore, and a low rocky islet on the west edge of this bank.

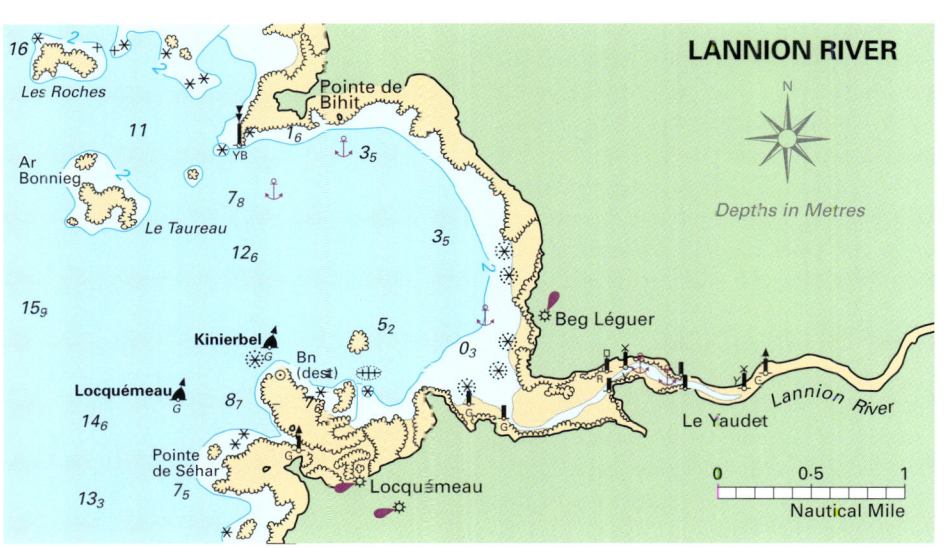

Lannion river

Once the islet is just abaft your port beam, steer to enter the narrows between the landing and the promontory at about 050°, leaving the moorings to port. Once through this neck, you'll see some more moorings to starboard, in a quiet bay off the south bank. You can anchor just clear of these moorings, in about 2½ metres at MLWS. The anchorage is sheltered from all quarters, but don't try to enter or leave the river in fresh onshore winds from between west and northwest.

ANSE DE LOCQUIREC

This shallow sandy bay forms the estuary of Le Douron River. It lies between Pointe de Locquirec and Pointe de Plestin, four miles southwest of the mouth of the Lannion River, and is sheltered from between west through south to southeast. At neaps you can tuck well in under Pointe de Locquirec, a couple of cables east of the harbour breakwater, but at springs you must

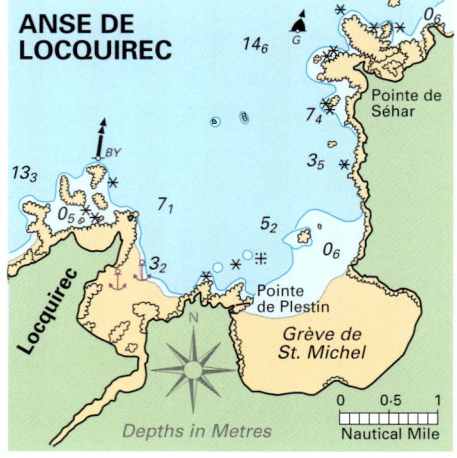

SECRET CORNERS OF THE MORLAIX ESTUARY

I've often noticed that there are plenty of cruising blind spots on the Brittany coast, where a higher level of navigational apprehension tends to increase the blinker effect as you negotiate a rocky entrance. I remember enjoying a few late-season days pottering around the Morlaix estuary, with a friend aboard who'd literally never been on a boat before. Fascinated by the sheer novelty of visiting new places by sea, she pored over the charts looking for harbours and anchorages, and her enthusiastic questioning led us to a fascinating 'blind spot' I'd forgotten existed. The Morlaix estuary is peppered with reefs and islets, through which three buoyed channels lead in towards the inner Rade. Even using instruments, navigators are usually preoccupied with identifying buoys and beacons and don't have much spare time for gazing around.

I was lining up for the Grand Chenal when several crucial marks vanished in a rain squall. While I was cursing the elements, our eager crew, entranced by the whole rocky prospect, was quizzing me about a low, mysterious-looking island that juts into the west side of the estuary opposite the rather faded resort of Carantec.

'That's Ile de Callot', I replied distractedly, trying to recapture the green Ricard beacon through the glasses. 'There's not much there, just a few houses and an old chapel.' But when I mentioned the tidal road across which the locals drive around low water, I knew we'd get no peace until we'd visited Callot and done a little exploring.

I also knew that for a foray ashore we'd be better off anchoring in the Penzé River, which joins the estuary on the west side of Callot. So we went round into the Penzé next morning after a night at anchor off Pen Lann. The Penzé River is narrower and more intricate than the Morlaix estuary and is itself a bit of a cruising blind spot. We fetched up opposite the gap between Ile de Callot and Carantec, anchoring on the edge of the river channel. Then we took the dinghy ashore to the south end of Callot where the tidal road was still well submerged. As we landed at a long stone jetty near the tidal road, two small fishing boats were cutting through Passe aux Moutons across the submerged road, which probably had five or six feet over the shallowest part. The tide was falling, but the fishermen knew almost subconsciously, from the look and feel of the estuary around them, that they had just enough water to slip through this passage now,

but wouldn't have even half an hour later.

Wandering along the sandy lane that follows the spine of Ile de Callot, we could see rocks, sand and mud appearing mysteriously on either side as the ebb fell away. As the water vanished, booted locals appeared with strange rakes and baskets ready to winkle out tasty morsels from the foreshore. Just beyond the narrowest part of the island, a tiny chapel stands on a mound surrounded by gorse. Huddled in a dip below it, a few weathered cottages look eastward over the beach. The chapel is built on the site of a much older church, started in 513 after a local Breton chief managed to oust an irritating Viking pirate who used Callot as a base from which to raid vessels and coastal villages around the estuary. Now this is one of the most peaceful corners of the Brittany coast. Peering across the estuary from this secret place, you may see yachts sailing through the Morlaix channels, but their crews certainly won't see you.

The Morlaix Estuary

anchor midway between the tip of Pointe de Locquirec and Pointe de Plestin to stay afloat. The river flows into the south corner of the bay, but is only navigable by small boats; yachts should keep well clear of its mouth. It's best to approach Locquirec above half-tide, when two drying rocks in the offing – Roc'h Parou (dries 1.6m) and Roc'h Felestec (dries 1.3m) – will be well covered. Various drying rocks extend up to a mile north and west of Pointe de Locquirec, so keep the bay well open as you come in.

ANSE DE TÉRÉNEZ

This narrow drying inlet lies on the east side of the Morlaix estuary and is approached via the Chenal de Tréguier above half-tide. Coming from seaward, follow the directions for the outer part of Chenal de Tréguier, but alter course to the southeast when Tourghi green beacon is two cables abeam to starboard. Steer towards the Annomer green spar beacon about half a mile away, leave it a cable to starboard and then head SSE towards Pointe de Térénez.

Near neaps, moderate draught boats can anchor and stay afloat close south or

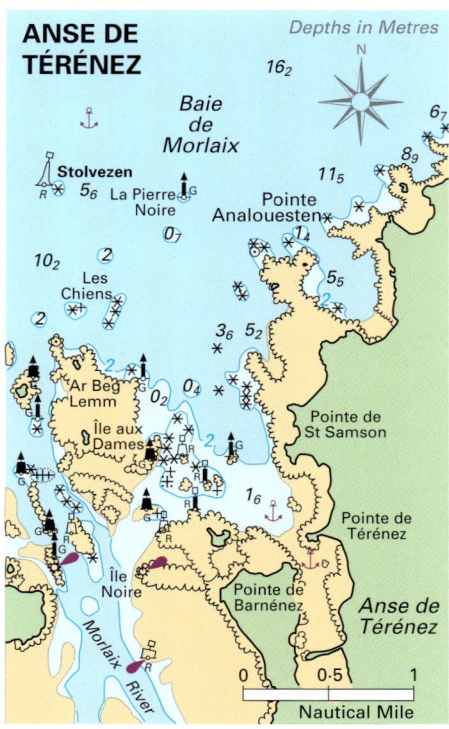

Morlaix estuary

SSW of Pointe de Térénez, protected from all winds except a fresh northwesterly. Bilge-keelers should be able to edge further into the inlet and take the ground. At springs, most keel boats will stay afloat about three cables northwest of Pointe de Térénez, and you can edge closer in between springs and neaps. There are a couple of convivial bistros ashore on the Térénez promontory.

PEN LANN

This prominent wooded headland forms the west arm of the Rade de Morlaix, jutting out whereic three entrance channels meet south of Ile Louet and Château du Taureau. In quiet or westerly weather you can anchor and stay afloat about 250 metres southeast of Pen Lann. Even in fresh onshore winds, you'll obtain some protection from the headland and by the natural breakwater

ANSE DE PAIMPOL TO THE PENZÉ RIVER

Anse de Térénez

Ile Louet and Château du Taureau

SECRET ANCHORAGES OF BRITTANY

Morlaix estuary at Locquénolé

of rocks and shoals in the outer reaches of the estuary.

Pen Lann is the usual holding anchorage for yachts waiting for the tide up to Morlaix, but it can also make a useful overnight anchorage, accessible at any tide and usually pretty snug unless the wind is fresh from the south. To reach this anchorage, enter the Morlaix River through either Chenal de Tréguier, the Grand Chenal or Chenal Ouest de Ricard, and then watch the echo-sounder carefully as you nudge inshore under Pen Lann. Near neaps, you can obtain greater shelter from onshore winds by edging further round to the south of Pen Lann.

The austere-looking fort on Ile du Taureau was built in the mid-16th century by the people of Morlaix, to protect the river against English marauders. In 1522 a sizeable English fleet had penetrated the estuary and sailed right up to Morlaix on the tide. After much hard fighting the English fleet withdrew with honour on both sides, but since then the Morlaix coat of arms has included a lion facing the English leopard, with the inscription: 'S'ils te mordent, mords-les' – If they bite you, bite them back!

CARANTEC

There's a pleasant fair weather anchorage on the northwest side of Pen Lann, with Grand Cochon green and white beacon tower bearing a shade south of east a couple of cables off. Coming from seaward above half-tide, follow the charted directions for the Chenal Ouest de Ricard, whose two white leading marks come into clear view as you draw abreast of Ile de Callot. Continue along this outer leading line at 187° until you are within 200 metres of Pierre de Carantec with its white-painted mark. Then edge to port a little to anchor just east of the leading line, a short 200m west of Basse Plate rock (dries 6.8m), which is prominent about a cable WNW of Grand Cochon beacon.

This spot has about half a metre at LAT, so can be used by moderate draught yachts even towards MLWS. However, quiet settled weather is needed, because the anchorage is much more open to swell than the inner part of the estuary at Pen Lann.

ANSE DE PAIMPOL TO THE PENZÉ RIVER

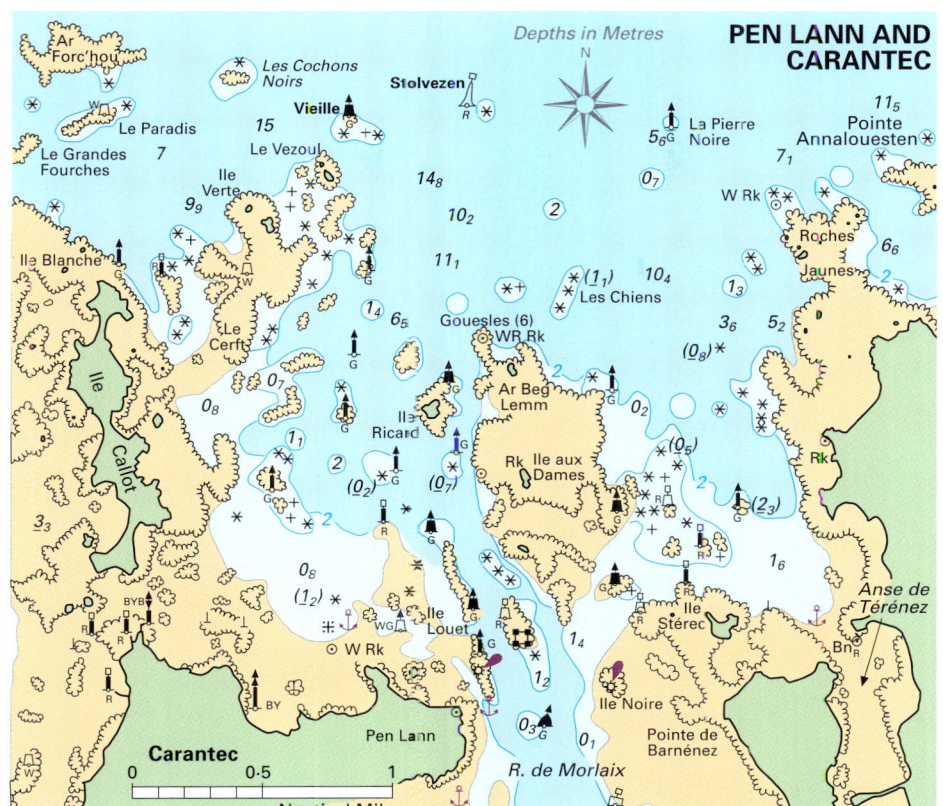

Penzé river

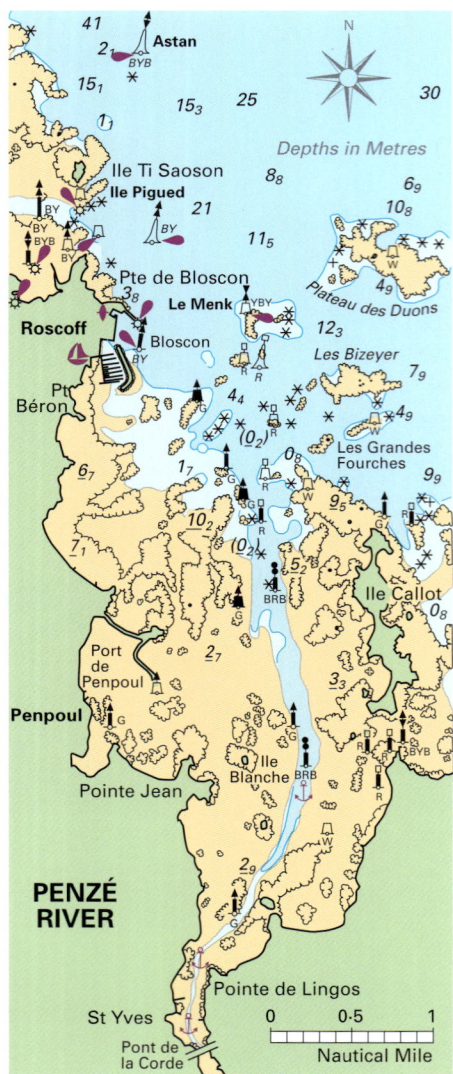

Stormalong at Penzé, the head of navigation on the Penzé river

THE PENZÉ RIVER

The Penzé estuary opens into the west side of the bay of Morlaix and is navigable for about four miles as far as St Yves. The river is quiet and sheltered and there are various anchorages in or near the narrow deep-water channel. The most protected is right up at St Yves, a cable southwest of L'Ingoz white beacon in about one and a half metres MLWS. A good spot lower down is more or less opposite the shallow Passe aux Moutons that separates Carantec from the south tip of Ile de Callot. If you fetch up in the main river channel about 200 metres south of Le Figuier spar beacon, it's not far to row ashore to Carantec at high water.

It's always preferable to enter the outer Penzé estuary above half-tide, starting from a position two cables west of Le Menk W Cardinal beacon tower and following the red and green beacon towers southwards.

USEFUL ADMIRALTY CHARTS

No 2026 Anse de Kernic to Ile Grande
No 2027 Ile Grande to Ile de Bréhat
No 2028 Ile de Bréhat to Plateau des Roches Douvres
No 2745 Baie de Morlaix, Ile de Batz to Pointe de Primel

USEFUL SHOM CHARTS

No. 7095 Baie de Morlaix – De l'île de Batz à la Pointe de Primel
No. 7124 Baie de Lannion De la Pointe de Primel à l'île Grande
No. 7125 Abords de Perros-Guirec – Les Sept Iles – De l'île Grande à l'île Balanec
No. 7126 De l'Ile Balanec aux Héaux-de-Bréhat – Cours du Jaudy
No. 7127 Abords de l'île de Bréhat – Anse de Paimpol – Entrée du Trieux

ONION JOHNNIES OF ROSCOFF

Some chefs wouldn't use any other type of onion, but nowadays you can't just 'Stop me and buy one'! Those splendid pink Roscoff onions were sold door-to-door in England for over 150 years, but are now more likely to be shipped over in containers to supermarkets than be seen swinging from the handlebars of an elderly bike ridden by an 'Onion Johnny'.

In the 14th century, some Roscoff onions were landed in Exeter from the *Sainte-Marie du Conquet* and thereafter the English developed a taste for this exquisite bulb with its sweet pink flesh. In 1828 a young Breton farmer, Henri Ollivier, started crossing the Channel regularly and selling his onions direct to cooks and housewives in southern England. Neighbours saw this was a profitable wheeze and by the 1920s over 1,400 boats were travelling between North Brittany and our own south coast.

The young men who actually knocked on doors, some only 12 or 13, usually worked for a bigger fish who shipped in strings of onions to convenient harbours. The onion sellers were called Johnnies by the English who thought that most Frenchmen were named Jean. They would eat onions three times a day, sleep in doorways under sacks of onions and work from dawn to dusk for three or four months a year.

Any Johnny returning to base with unsold strings would have his meagre wages docked. Carrying up to 100kg on long poles, the young Johnnies would persuade housewives with Gallic charm and seductive broken English. They were well known in southern England and in Wales. Some ventured as far north as Edinburgh and even the Shetlands.

After World War II fewer Johnnies crossed the Channel, but I remember seeing them in South Wales markets during the 1980s. Pink Roscoff onions are now sold in up-market greengrocers, but far better to buy a couple of strings in Roscoff and bring them home on the boat.

They are perfect for making classic French onion soup, finely chopped and sautéed to a syrupy golden brown before adding the water. Sliced in very fine rings, pink Roscoff onions go well with a simple tomato salad, lightly dressed with *vinaigrette*.

The Chenal de Batz seen from the island

CHAPTER 3

ILE DE BATZ TO L'ABER-ILDUT

As you work westwards past Roscoff and Ile de Batz, the north coast of Brittany becomes more low-lying and rather tricky to identify from offshore. It takes on a harder, more austere mood than hitherto, and there are plenty of drying rocks that don't appear until the ebb is well away. You can reckon a good 35 miles from the Morlaix estuary to l'Aberwrac'h, a full tide under sail, and most yachts aim to make a quick passage between the two without lingering on the way. This doesn't seem, on the face of it, an area suitable for nudging into secluded bays and inlets. Yet even this stretch of coast has a handful of natural havens where, given the right conditions and a favourable forecast, you can tuck close in and lie to your own ground-tackle undisturbed.

There are several attractive spots off the southeast side of Ile de Batz, in the narrow Chenal de Batz that only just seems to separate the island from the mainland. In fact there are now clusters of moorings in some of the positions where one used to lie at anchor. This is a pity in a way, since perfect seclusion is now that bit more elusive, but on the whole the moorings are a welcome addition. The tidal streams can be strong through the Chenal de Batz and there are various unpredictable eddies close inshore where, on a long scope near low water, you can find yourself swinging rather closer to dry land than intended.

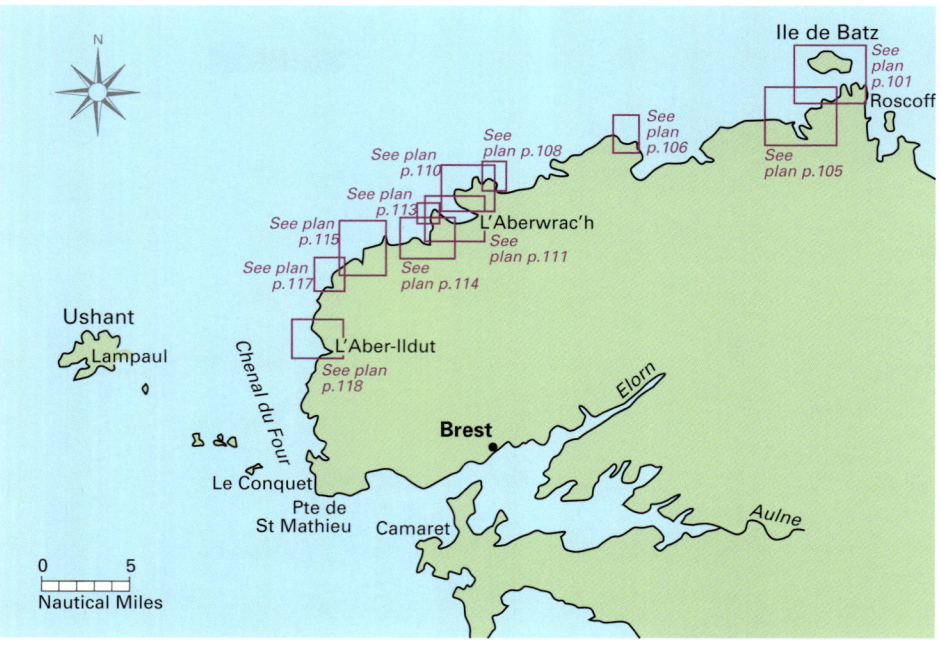

The Chenal de Batz, usually called the Canal de l'Ile de Batz on English charts, is a fascinating narrow passage just over three miles from end to end. Most yachts use it as a short cut when coasting, rather than make a long detour round the rocky dangers seaward of Ile de Batz. The channel looks intricate in prospect, but is well marked and fairly straightforward once you get going. The east end is shallow, with soundings down to half a metre at LAT, so you really need a couple of hours' rise of tide at springs. This is usually no problem when carrying a fair stream in either direction between the Morlaix estuary and l'Aberwrac'h, but needs watching if you are making shorter local hops between anchorages.

The western end of the Chenal de Batz, near Basse Plate N Cardinal beacon tower, passes about one and a half miles off the mainland to clear a wide area of dangers extending well offshore. This whole plateau of coastal reefs needs to be skirted by a safe margin as you curve round to the southwest to enter the bay which has Ile de Siec on its north side and tiny Moguériec harbour in its southwest corner. The anchorages in this bay are attractive and secluded places to lie for a day or two, but they can also be useful passage stops in quiet or offshore weather, cutting the distance along the coast to l'Aberwrac'h entrance to something like 25 miles. This leg makes a comfortable one-tide passage that you can usually sail, even in a head wind, without much risk of picking up a foul stream before you arrive.

The anchorage off Ile de Siec, between Querelevran rock and the west end of the island, is a good spot in easterlies, especially near neaps. Stormalong once lay here for several days late in the season while a near gale from more or less due east raged in the English Channel. We had beat along from l'Aberwrac'h in a nasty steep chop while the wind was freshening, running out of fair tide before reaching the Chenal de Batz and grateful for the lee of Ile de Siec and its nearby crook of mainland. I had intended to slip round to Morlaix on the next tide, but the wind became so chill and malevolent that we laid out a second anchor, lit the cabin stove and settled down to wait, reading good books between long, leisurely meals. Each afternoon we would row across to the island for a bracing walk to work up an appetite for dinner.

You can land with the dinghy at an old stone jetty just northeast of the anchorage, opposite the prominent wartime ruins of a house at the west tip of Ile de Siec. The island has one or two holiday cottages, but is otherwise uninhabited. At low water, you can walk across a sand-spit causeway to the mainland and the small hamlet of Dossen. An alternative anchorage in winds with any south in them is off the entrance to Moguériec, on the south side of the bay. This is a snug spot in offshore weather, with the advantages of a couple of shops and a congenial hotel-restaurant ashore. In the French holiday season, you can enjoy a shower at the campsite just inland from the south quay.

Approaching Portz Kernoch on Ile de Batz

L'Aber-Ildut

The soothing upper estuary at L'Aberwrac'h

Some 10 miles west of Moguériec and Ile de Siec, on the west side of the shallow Anse de Kernic, is the small natural harbour of Pontusval. This U-shaped inlet in the low-lying coast is entered from northward between two wide areas of drying rocks that straggle out from each side of the mouth. These rocks help protect the inlet from swell so that, in moderate or offshore weather, a yacht can anchor and stay afloat in the narrow approach channel to the drying part of the harbour.

While hurrying up-Channel, we've sometimes used Pontusval after coming up through the Chenal du Four early on the tide and made sufficiently good progress round the corner to carry the east-going stream beyond l'Aberwrac'h entrance. Starting next day from Pontusval on the last of the west-going stream, you can aim to get through the Chenal de Batz and along the coast as far as Ploumanac'h before the tide turns foul again.

The North Brittany coast around Pontusval often feels rather hostile and one can be reluctant to turn inshore to find the narrow entrance. However, the approach is actually quite simple in practice, so long as you start from the Pontusval E Cardinal buoy, moored just opposite the entrance not quite a mile offshore. The tall radio aerial a couple of miles inland is a useful mark when trying to locate this buoy.

Heading west towards l'Aberwrac'h, you'd normally pass two miles north of the tall lighthouse on Ile Vierge, on the leg between Lizen Ven W Cardinal buoy and a turning waypoint a mile or so north of Le Libenter W Cardinal buoy at the entrance to the estuary. However, on a quiet day when the tide is right, an intrepid navigator can edge in close to Ile Vierge from Lizen Ven buoy, leave the west end of the island close to port, and creep gently southeast between the reefs into a narrow sandy inlet known as Portz Malo. This natural drying harbour is used by a few local fishing and seaweed-gathering boats and offers a splendid hideaway for a yacht with bilge-keels. You can lie here in perfect solitude in the lee of Ile Vénan, less than two miles from the often crowded marina at l'Aberwrac'h.

You need to keep an eye on the forecast, of course, but Portz Malo is worth the slight stress of pilotage involved in getting in. My most memorable visit to this secret place was in a Westerly Konsort, an ideal bilge-keeler for this kind of exploration. The weather was set fair for a while with a light breeze from the southwest. It was somewhat eerie to be dried out on firm sand at midnight a mile inside Ile Vierge, watching its single powerful beam sweep this potentially treacherous corner of North Brittany. We weren't disturbed by anyone until late morning, when a Frenchman staying on holiday nearby knocked diffidently on the hull and presented us with a bag of freshly gathered mussels, just in time for lunch.

On then to the real heart of the Côte des Abers – l'Aberwrac'h, l'Aber Benoît, L'Aber-Ildut – and a selection of rarely visited, off-beat anchorages in the northern approaches to the Chenal du Four. This northwest corner of Brittany can be an absorbing cruising area in its own right, and yet most yachts pass through quickly, perhaps calling at l'Aberwrac'h for a night before pressing on down through the Chenal du Four bound for Camaret, the Raz de Sein and the Biscay coast.

The Chenal du Four is an important tidal gate for this corner and you need to carry a fair stream when cruising anywhere between l'Aberwrac'h and Pointe de St Mathieu, where the Chenal du Four joins the spacious reaches of the Brest estuary. Yachts heading south from l'Aberwrac'h right through the Four would normally aim to stem the last of the east-going tide on the north coast, leaving l'Aberwrac'h entrance about an hour before local high water. There is more flexibility, however, if you are making shorter hops between anchorages.

A fresh wind over tide kicks up a steep chop in the southern, shallowest part of the Chenal du Four. Poor visibility is common in the whole area, although the buoyage is excellent and it's usually not difficult to pick your way around even in quite murky conditions. The real potential menace is heavy northwesterly swell, especially at the north end of the Chenal du Four, which can soon make some of the smaller anchorages untenable.

ILE DE BATZ

There are several possible anchorages in the Chenal de Batz (or Canal de l'Ile de Batz), that fascinating shallow sound separating the island from the Brittany mainland. Working from the east, the first spot you reach is between Ile Pighet white beacon and Duslen S Cardinal beacon tower, but closer to Ile Pighet than Duslen to avoid the drying rock on the direct line between the two. This anchorage area, which now has several moorings, is best around neaps if you prefer to tuck well up to the north out of

Anchorage at Portz Kernoch

Ile de Batz lighthouse

the strongest part of the tidal stream. Although this reach feels rather exposed, it's actually quite sheltered from between northwest and southwest, especially as the tide falls away.

Porz an Ilis Tucked snugly under the southeast corner of Ile de Batz, the drying bay known as Porz an Ilis is used only by a few local fishing boats and we have sometimes anchored comfortably

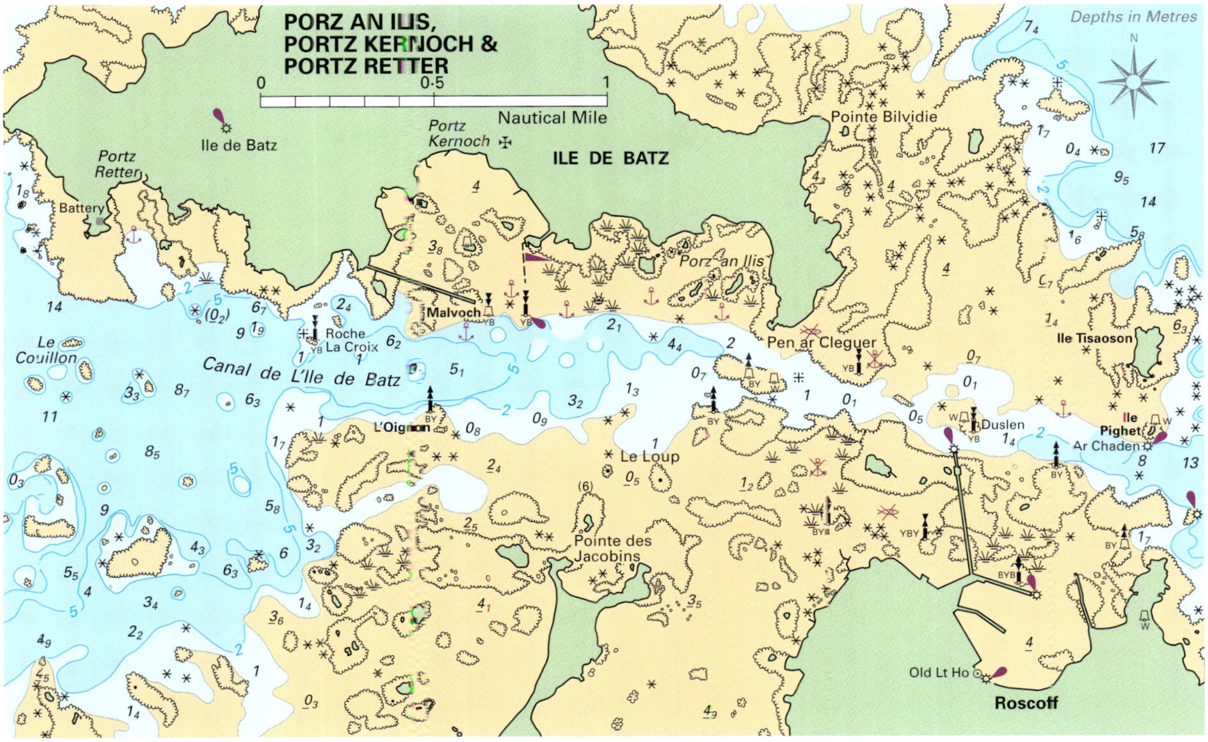

ILE DE BATZ TO L'ABER-ILDUT

PORZ AN ILIS, PORTZ KERNOCH & PORTZ RETTER

here at neaps. Coming from the east through the Chenal de Batz, the channel leads more or less midway between Per Roch N Cardinal beacon tower and the southeast tip of Batz – Pen ar Cleguer. Once west of Pen ar Cleguer, you can nudge north carefully at neaps towards Porz an Ilis, anchoring just south of one of the gaps between the various islets in the bay. Watch the echo-sounder and work out your low water depth carefully before settling down at anchor. The main danger to avoid is an isolated rock (drying 2.1m) that lies not quite 250 metres southeast of Ile Aukuint and about 400 metres west of Pen ar Cleguer.

Off Portz Kernoch A little further west along the Chenal de Batz, you can fetch up 1½–2 cables east of the S Cardinal spar beacon that marks the south extremity of the long low water landing slip at Portz Kernoch. At neaps

Ile de Batz church

Malvoch N cardinal beacon tower

SECRET ANCHORAGES OF BRITTANY   101

Portz Kernoch

Portz Retter

you can edge a little inshore to find quieter water, but make sure that the landing slip beacon is bearing no less than about 260° as there are ledges of drying rocks to the north of this line.

There's also an anchorage just off Portz Kernoch entrance, within a cable WSW of Malvoch S Cardinal beacon tower. Both these spots are pretty well sheltered in any northerly winds, although a tidal scend is apt to roll through the Chenal de Batz on the flood.

Just into Portz Kernoch At dead neaps, moderate-draught yachts can usually stay afloat just into Portz Kernoch itself, opposite the entrance and about halfway along the low water landing slip. Ile de Batz is picturesque and well worth an expedition ashore, but don't leave your boat unattended in this harbour anchorage, because the island launches are always coming and going and you may need to move.

Portz Retter This interesting, little known anchorage off the southwest corner of Ile de Batz is not shown clearly on the latest Admiralty 2745, which rather glosses over Portz Retter with a few vague, cost-conscious indications of reefs. However, the intricate engraving of my venerable black and white fathoms chart of Ile de Batz clearly shows all the protecting fingers of rock around this pronounced bay where several local boats have moorings. Portz Retter is protected not just by its west promontory (where the metric chart shows the old battery) but by a long plateau of rocks extending 300 metres further south. This would have been a useful anchorage for islanders in the days of sail, close to the open sea and quite easy to get into despite the discouraging fuzz on the modern chart.

The bay is divided into two by a rocky spur, and the western half is where the local boats lie. This inlet is usually easiest to enter at low water, approaching from due south having reached a position more or less midway between Roche La Croix S Cardinal beacon and Basse Plate N Cardinal tower. You come in with the battery point fine on your port bow and Ile de Batz main lighthouse on your starboard bow. Steer to leave the most southerly rocky outcrop on the *east* side of the inlet fairly close to starboard, which helps you avoid the outer (and trickier) drying rock on the west side. Then nudge in carefully at just west of north, steering towards an old lifeboat slip at the head of the bay. Watch the echo-sounder and fetch up just outside the local moorings.

I anchored in Portz Retter many years ago to escape a strong northeasterly, lying perfectly snug while massive seas battered the weather side of the island less than half a mile away.

ILE DE SIEC AND MOGUÉRIEC

Ile de Siec lies very close to the North Brittany coast, a little over two miles SSW of the west entrance to the Canal de l'Ile de Batz. Numerous drying rocks extend up to one and a half miles offshore between Roscoff and Ile de Siec, but there's an intriguing and rather remote deep-water anchorage about a cable southwest of the west tip of Ile de Siec, sheltered in moderate weather from northeast through east to south.

Moguériec's sleepy outer harbour

CHOOSING YOUR FISH IN FRENCH

What's the real difference between *morue* and *cabillaud*, *bar* and *barbue*, and should it be *merlu* or *merlans*? Finding your way around a restaurant menu is one thing, but faced with the glistening array of a Brittany fish stall it's hard to know what to choose or how to order it.

Turbot, sardines and sole are easy, as they have the same name in French and English. *Saumon* is not far removed from salmon, *anchois* from anchovies or *truite* from trout. Delicious John Dory are *St Pierre* in French, not to be confused with *Coquilles St Jacques* – mouth-watering scallops. Personally I would avoid *anguille* as I'm not keen on eels, and definitely steer clear of *brochet* as the best thing to do with an oven-baked pike is throw away the fish and eat the stuffing.

Bar is the noble bass and *barbue* the fleshy succulent brill, both important names to remember. *Cabillaud* is cod but *morue* is generally salt cod, an acquired taste best sampled in a restaurant where someone else is doing the work. *Lieu* is saithe, a rather anaemic, cardboardy sort of fish not worth buying unless you have a ship's cat to feed. *Lieu noir* is coley (usually rather dismal) and *lieu jaune* is pollock (often very fleshy and tasty), but either can turn up as *poisson du jour* on a menu, labelled simply as *lieu* if labelled at all. Hake is generally called *merlu*, although it sometimes turns up as *colin*. *Merlans* is whiting and plaice is either the masculine *le carrelet* or the feminine *la plie*. The fish *julienne* in France is ling, occasionally called *morue longue* – tasty if prepared *à la Provençale*. *Lotte* is always worth buying in markets or ordering in restaurants. This is the lugubrious looking monkfish (sometimes called angler fish in English!) whose flesh is firm yet moist and juicy, not unlike prime chunks of large prawn. My favourite *lotte* recipe is *Gigot de Lotte à l'Ail Rose*, a whole monkfish tail roasted in the oven with pink garlic cloves and tomatoes. The fish is carved at the table like a leg of lamb.

When a magnificent *plateau de fruits de mer* is brought to the table the *crustacés* (shellfish) will include *tourteau* (the familiar flat-backed crab) or *araignée* (spider crab – fiddly but luscious and sweet). You'll have tiny *crevettes grises*, the tasty grey shrimps that you crunch whole; succulent *crevettes roses* (those reliable red prawns); and usually *bulots*, large, meaty whelks that the French do so well and the English ruin with vinegar.

There'll always be plenty of *bigorneaux*, small black winkles that are much more enjoyable than they look, and you may have half a dozen *huîtres* (oysters). If you have splashed out on a *plateau royale*, an aristocratic looking *homard* (lobster) will hold pride of place in the centre of the platter. Most cruising crews are familiar with *moules marinières* or *moules au cidre*, but less common are the elegant *palourdes*, delicately flavoured clams sometimes served in a similar way to mussels, but more often as the scrumptious *palourdes farcies* – lightly grilled with garlic and herb butter.

Finally, never ask for *macareux grillés* if you fancy grilled mackerel, as even the French don't eat grilled puffin these days! Try ordering *maquereaux* instead.

ILE DE BATZ TO L'ABER-ILDUT

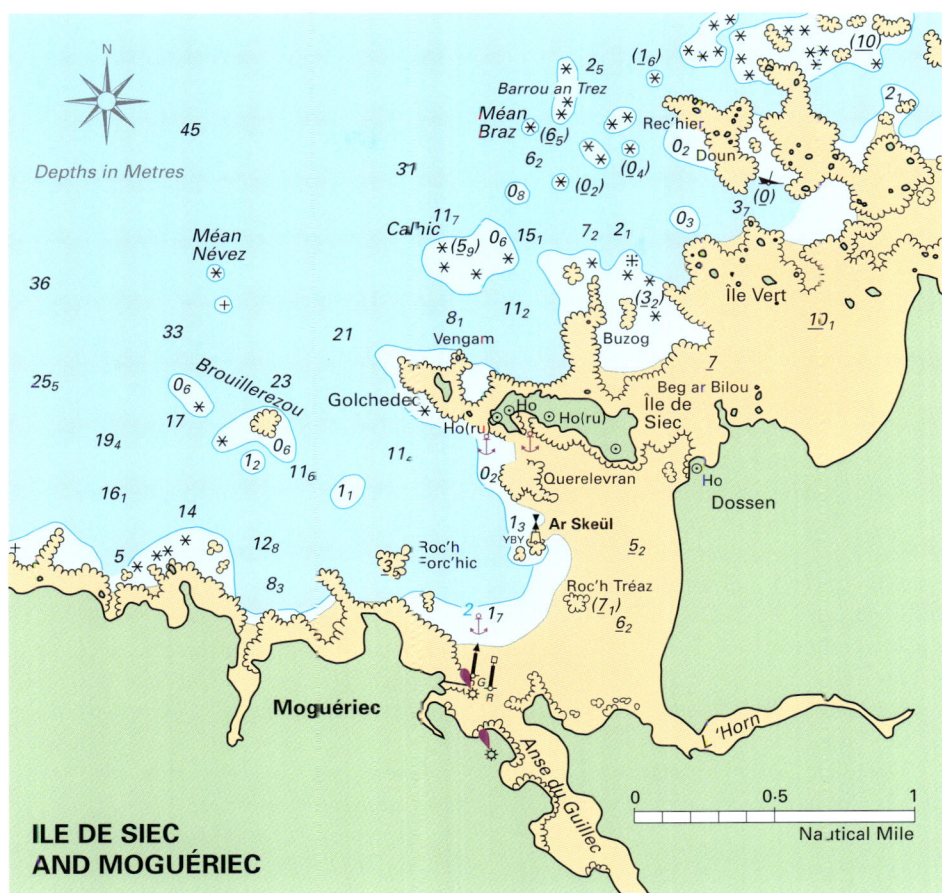

ILE DE SIEC AND MOGUÉRIEC

Approach from the NNW within two hours of high water, from a position two and a quarter miles due west of Basse Plate N Cardinal beacon tower. Make good 162° towards Moguériec pierhead (three miles away), aiming to pass a quarter of a mile west of Golchedec – the prominent islet close off the west tip of Ile de Siec. There are dangers to the west of this approach line, of which the most significant is Méan Névez (dries 3.3m).

Once the west tip of Ile de Siec bears north of east, come to port and head more or less east into the bay formed by Golchedec, Ile de Siec and Querelevran rock. A good spot to anchor is with the west tip of Ile de Siec bearing north true, with the top of Querelevran a cable to the southeast. At neaps, in quiet weather, you can edge further east between Ile de Siec and Querelevran, but be sure to anchor in the sandy channel, clear of the rocky ledges to the north and south.

Mogueriec

SECRET ANCHORAGES OF BRITTANY 105

Pontusval near high water

The small fishing village of Moguériec lies at the mouth of the Rivière du Guillec, a mile south by west from Ile de Siec. A stone pier protects its local moorings from the north. Towards neaps, in quiet or offshore weather, you can anchor and stay afloat a couple of cables NNE of Moguériec pierhead, sometimes a bit closer in if you have a modest draught.

PONTUSVAL

This tiny natural harbour, hemmed in by rocks, lies 12 miles WSW along the coast from Ile de Batz and about 10 miles ENE from Ile Vierge lighthouse. Few yachts call at Pontusval, perhaps because the coastline either side often looks rather inhospitable. However, the approach is fairly straightforward (in daylight and preferably above half-tide) given reasonable visibility, light or offshore winds, and an absence of swell.

The most useful landmarks are Pointe de Beg Pol lighthouse, a mile west of the

entrance, and a tall radio mast a couple of miles inland from the lighthouse. You will find a surfeit of water towers, so be careful about using these for establishing your position. The pilot books give various transits, but the easiest guide is to reach Pontusval E Cardinal buoy first and then make good 178° towards the entrance (less than a mile away), allowing for the cross-tide which you assessed at the buoy.

On entry, leave Pecher green buoy to starboard, An Neudenn red beacon tower to port and three white-painted rocks to starboard. At springs, keel boats should fetch up opposite the second white rock; at neaps you can venture past the third white rock, anchoring in mid-channel between this and Kinloc'h du Dédans spar beacon. Pontusval provides good shelter in any winds from the south.

CORRÉJOU

This small, rather off-beat Brittany haven lies not quite three miles east of Ile Vierge and is rarely visited by cruising yachts, probably because of its discouraging expanse of dangers extending a good two miles offshore. If, however, you approach Corréjou from the west (i.e. from the direction of l'Aberwrac'h), the pilotage is not as

SECRET LANDINGS OF THE SECOND WORLD WAR

Even the most placid inlets and estuaries have stories to tell. Cruising boats heading for Biscay often call at L'Aberwrac'h on the northwest corner of Brittany, a handy staging post for the Chenal du Four and the next leg south towards the Raz de Sein. The *abers* on this remote fringe of Finistère have a special atmosphere steeped in the tang of ozone, seaweed and shellfish. Intricate patterns of reefs and small islands change continuously with the tides, creating hideaways for local boats and visitors who like anchoring in secret places. L'Aberwrac'h looks an unlikely haven from offshore and there seems no way in through the rocky chaos until you spot the Libenter buoy with its doleful whistle. Then a smaller red can leads to a pair of beacon towers and you are gradually enclosed without really noticing. Following this wild inlet towards the snug upper reaches, you pass a grey stone fort on Ile Cézon, which was manned by German gunners during the Second World War when the occupying forces arrived in June 1940. Two years later this fort became a tiny brick in Hitler's grandiose Atlantic Wall. During the winter of 1943, though, under the very nose of Cézon, there began one of the most intriguing nautical episodes of the Hitler war, when the French Resistance worked undercover with the Royal Navy to repatriate British airmen shot down during the heavy and sustained Allied bombing raids on Brest. Ditched aircrews who avoided capture were sheltered by the Resistance and stealthily moved north by night to some lonely house on Presqu'île Sainte-Marguerite, between L'Aberwrac'h and its neighbouring inlet L'Aber Benoît. The pick-ups were usually made from Ile Tariec, half a mile west of Sainte-Marguerite off the mouth of L'Aber Benoît. At low springs you can walk out to Tariec across the sand and escapees were taken to the island on carts used by local seaweed gatherers, hidden under piles of wet slimy wrack. Then they'd lie low until a Motor Gun Boat arrived from England on the next moonless night.

Anyone who has cruised this coast will understand the navigational skill involved in these operations. I've often wondered how MGB navigators found this spot in pitch darkness with no instruments except compass and echo-sounder, on a coast where powerful tides sluice through off-lying reefs. The lighthouses were out and it must have been extremely tense approaching this formidable lion's mouth on blind dead-reckoning after a 100-mile Channel crossing. The rescuers would probably have come in from due north near high water, heading for Ile de Rosservo and calculating from soundings over safely covered outer shoals when they were near latitude 48°36'. Then they'd have turned due east towards Ile Guénioc, creeping round its southern edge in the black night. Anchoring with muffled cable off the southeast side of Guénioc, the skipper would have sent a boat ashore to pick up the waiting airmen from Tariec, hoping to be out again and clear of the coast while it was still dark.

L'Aberwrac'h estuary

tricky as it seems from a first glance at Admiralty 3668. Quite a wide corridor – the Chenal Occidental – cuts close inshore between Plateau du Lizen Ven and the mainland.

The simplest way to visit Corréjou is by leaving l'Aberwrac'h through Chenal de la Malouine at local high water, where you emerge into more or less open water one and a half miles due west of Ile Vierge lighthouse. At high water l'Aberwrac'h you'll have a bit of room to spare out through the Malouine and the coastal stream will be trickling east for another hour or two, which is just what you want for Corréjou.

Head seawards until Ile Vierge lighthouse bears just south of east and then turn east along the coast, steering to leave the lighthouse four cables to starboard. Drawing abreast of Ile Vierge, you should see a red can buoy on your port bow, which marks the outer northwest corner of the Chenal Occidental. When you are on the line between this buoy and Ile Vierge lighthouse, or in any case when the lighthouse bears 210° half a mile off, turn a little south of east to follow a track leading about three cables seaward of the next two mainland promontories. The second promontory is your target, with its prominent chapel of St Michel Noblet, and you should steer down towards this low headland until Ile Vierge lighthouse bears exactly 270° astern.

Now steer due east true, keeping Ile Vierge lighthouse astern at due west true, carrying on past the St Michel headland for about three-quarters of a mile until you have just passed two distinctive above-water rocks that both have pronounced heads – Carrec Crom and Petit Crom. These two rocks lie about half a mile northeast of the tiny harbour island of Corréjou and you have to skirt east-about them before coming back in towards the anchorage. So continue just east of Petit Crom and then turn due south to leave it fairly close to starboard, carry on due south for another quarter of a mile and then steer about 220° towards the harbour area. Anchor a couple of cables east of the slip towards springs, but calculate your low water depth carefully before the ebb runs too far away. At neaps you can edge further in. Keep clear of the local fishing boat moorings and buoy the anchor, just in case.

PORTZ MALO (ILE VIERGE)

A fascinating spot for bilge-keelers, this narrow sandy inlet lies amongst the various rocks and islets between Ile Vierge and the mainland. It should only be visited in quiet settled weather and preferably near neaps, but the approach is easier than you might imagine, starting from a position three-quarters of a mile northwest of Ile Vierge lighthouse. You need to identify Ile Valan, a small island lying two cables south of the west tip of Ile Vierge; Ile Valan has a detached rock, 15m high, close northwest of it. Also identify Lanvaon lighthouse, a white square tower a couple of miles inland behind Ile Vierge to the southeast. Lanvaon normally forms the rear leading mark for the Grand Chenal de l'Aberwrac'h.

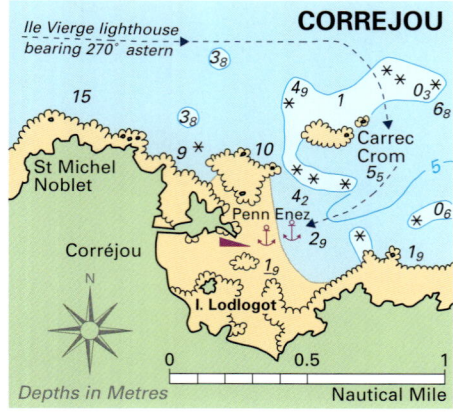

ILE DE BATZ TO L'ABER-ILDUT

Traditional seaweed boat on the beach at Corréjou

Bring Ile Valan open its own width to the west of Lanvaon lighthouse, with the latter bearing 137°, and steer in between Ile Vierge and Ile Valan on this transit. Continue towards Lanvaon until Ile Valan is abeam about 80 metres to starboard and then come to port a shade so as to leave Ile Vénan (19m high) not quite 100 metres to port and the much smaller Enez ar Vir (9m high) about a cable to starboard.

SECRET ANCHORAGES OF BRITTANY   109

ILE DE BATZ TO L'ABER-ILDUT

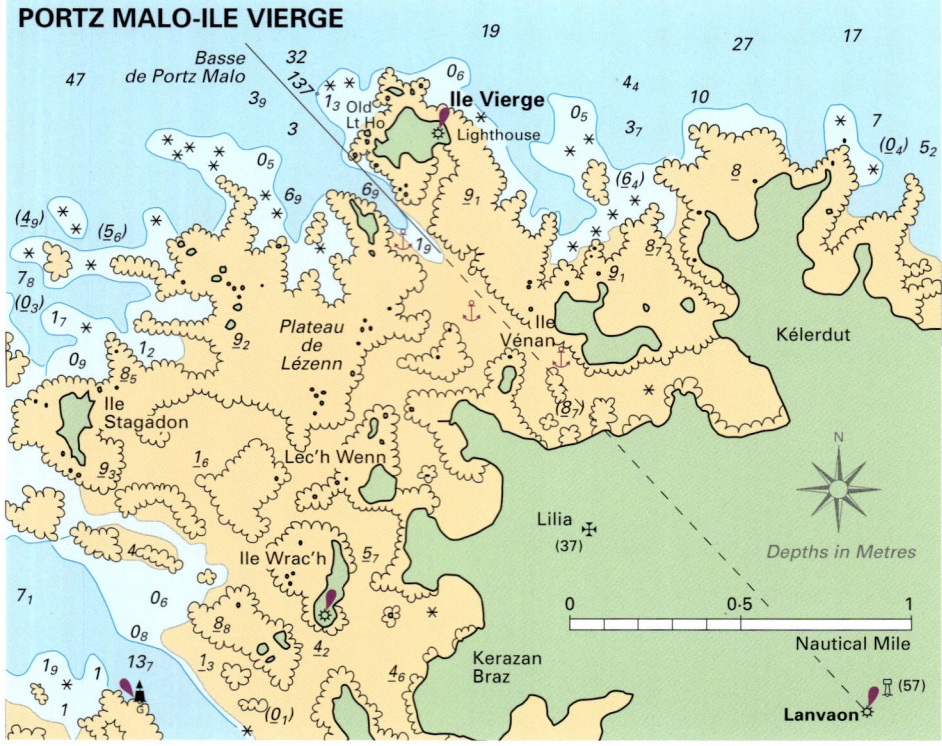

Ile Vierge lighthouse

Fetch up off the south side of Ile Vénan, where the sandy channel dries between 3½–5 metres LAT. There are one or two patches of rock and weed that need to be avoided as you take the ground. Portz Malo is sheltered in any winds from the south and is probably the least visited hideaway on the North Brittany coast.

L'ABERWRAC'H ESTUARY

Of the many yachts that visit L'Aberwrac'h each season, the majority are hurrying either home or south and simply stay overnight at the comfy marina at La Palue or lie to one of the club moorings. However, there are one or two anchorages around L'Aberwrac'h that allow you to get away from the madding crowd.

In settled weather, especially from the south, you can fetch up in the outer part of the estuary, 1½–2 cables ESE of the two Iles de la Croix and just about a cable SSW of Basse de la Croix green conical buoy. Approach the islands from the ESE and anchor well clear of a small rock (dries 3.4m) that lies just over a cable east of the north end of the larger island. On the other hand, don't stray too far south towards the rocky patches lurking up to a quarter of a mile north of Penn Enez, the tip of Presqu'île St Marguerite.

There are several sheltered anchorages above Pointe Cameuleut in the upper reaches of the river, but you must keep clear of local moorings and various mussel beds along the foreshore, the latter marked by patches of withies. There's a good spot half a mile or so above Pointe Cameuleut off the south bank, but it's really a question of finding swinging room on the day. The pool up at Paluden provides a pleasant anchorage, protected from all quarters and with about one and a half metres depth at MLWS 100 metres southeast of Paluden quay. There's a good family restaurant nearby and the small town of Lannilis is just over a mile walk to the south.

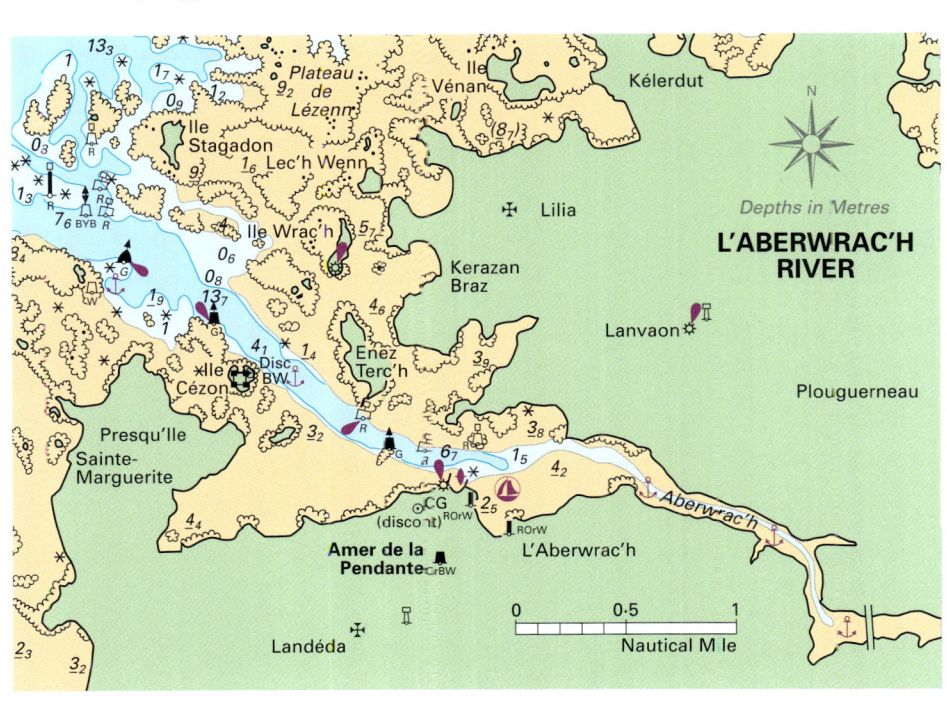

GOÉMONIERS OF THE ABERS

Cruising boats heading for Biscay often call at L'Aberwrac'h on the northwest corner of Brittany, a useful staging post for the Chenal du Four that leads south towards Brest. As well as being a handy haven, L'Aberwrac'h is also one of my favourite ports of call, with its special estuary atmosphere steeped in the tang of ozone, seaweed and shellfish. Intricate patterns of reefs and small islands change as the tide creeps in and out, making snug hideaways for moored local boats.

But there are two other 'Abers' on this remote fringe of Finistère – L'Aber Benoît and L'Aber-ildut. As you approach any of these rather wild, distinctly Breton inlets, you'll often see one or two long low boats working among the outer rocks, each equipped with what looks like a builder's crane on board. As you watch, a hydraulic grab will swing out to one side, lower towards the water and gather up great mechanical armfuls of seaweed from rocks at or just below the surface. As their grabs take the weight the boats lurch alarmingly, but steadily, load by load, huge piles of seaweed are collected on board. These boats are the *goémoniers* – the seaweed gatherers – and the iodine-rich wrack harvested around these coasts seems to have dozens of amazing commercial uses.

In the old days it was a much slower job harvesting seaweed using flat-bottomed sailing boats. These traditional *goémonier's* would sail slowly along the coast for two or three hours each side of high water while their crew used long-handled rakes to gather up the seaweed and pile it into the hold. As the tide started ebbing the boats would come inshore to a sandy beach where they would dry out. As low water approached, the land crew would bring horses and carts down to the beach, which were loaded alongside the dried out boats and driven back ashore.

Goémoniers were found all around the Brittany coast, but especially in the northwest around the Abers and among the islands opposite the Chenal du Four. Seaweed was landed at L'Aber-ildut, L'Aber Benoît, L'Aberwrac'h and also at tiny Port du Corréjou (sometimes spelt Koréjou), which lies a couple of miles east of Ile Vierge lighthouse. At Plouguerneau, not far inland from Koréjou, the fascinating Musée des Algues shows the numerous different types of local seaweed and how they were gathered and used. You can see an old *goémonier's* boat and visit the reconstruction of a *goémonier's* house. On a grassy headland behind Corréjou harbour, you can still see traditional stone-lined ovens where seaweed was burned to break it down into more portable ash. These long narrow troughs were filled with piles of dried weed, which were covered over with more stones and set alight. The ash required fewer carts for its transportation inland to iodine-extracting plants. Seaweed was, and still is, used as fertiliser in Brittany, being spread straight onto the land and left to dry out and partially decompose before being dug in.

The present harbour at Corréjou has grown around the site of the old seaweed gatherers' village of Trolouc'h, which is now a part of the Ecomusée des Goémoniers et de l'Algue. There are plans to reconstruct the village to conserve and display what represents an important part of the history and traditions of the area. Modern seaweed gatherers still use Corréjou for unloading their slippery cargoes, but today there are many sophisticated markets for seaweed, especially in the pharmaceutical industry, for cosmetics and as natural ingredients in herbal remedies. The seaweed gathering continues and the *goémoniers* live on.

Goémonier statue at Corréjou

ILE GUENIOC

This rather exposed anchorage in the approaches to L'Aber Benoît is included as a historical curiosity rather than a practical haven. It is here, or hereabouts, that naval MGBs (Motor Gun Boats) from South Devon or Cornwall used to anchor in the dead of night during the Second World War to pick up British airmen who had been shot down during the bombing raids on Brest and sheltered by the French Resistance.

To cross 100 miles of English Channel and then find this place without instruments on the rocky, tide-swept and, at that time, completely unlit Brittany coast was an extraordinary feat of navigation. Nowadays we can come in confidently using GPS to find the Libenter and Petite Fourche W Cardinal buoys even in murky visibility. From La Petite Fourche you head just east of south to pick up Ruzven Est green buoy and then continue to pass 300 metres west of Ile Guénioc. Now make for the red spar beacon that stands south of the island, steer to leave it a short 100 metres to port, curve east for 300 metres beyond the beacon and then turn due north to anchor a cable off the southeast edge of Ile Guénioc. A piece of cake on a fine summer day with all your instruments, but not so easy on a black winter night in 1943.

This is not an anchorage for staying overnight, but it's fascinating to lie here over a spring low water on a quiet afternoon, sheltered by exposed reefs and musing on a very different era that unfolded not so very long ago.

L'Aberwrac'h outer estuary

Petit Pot de Beurre

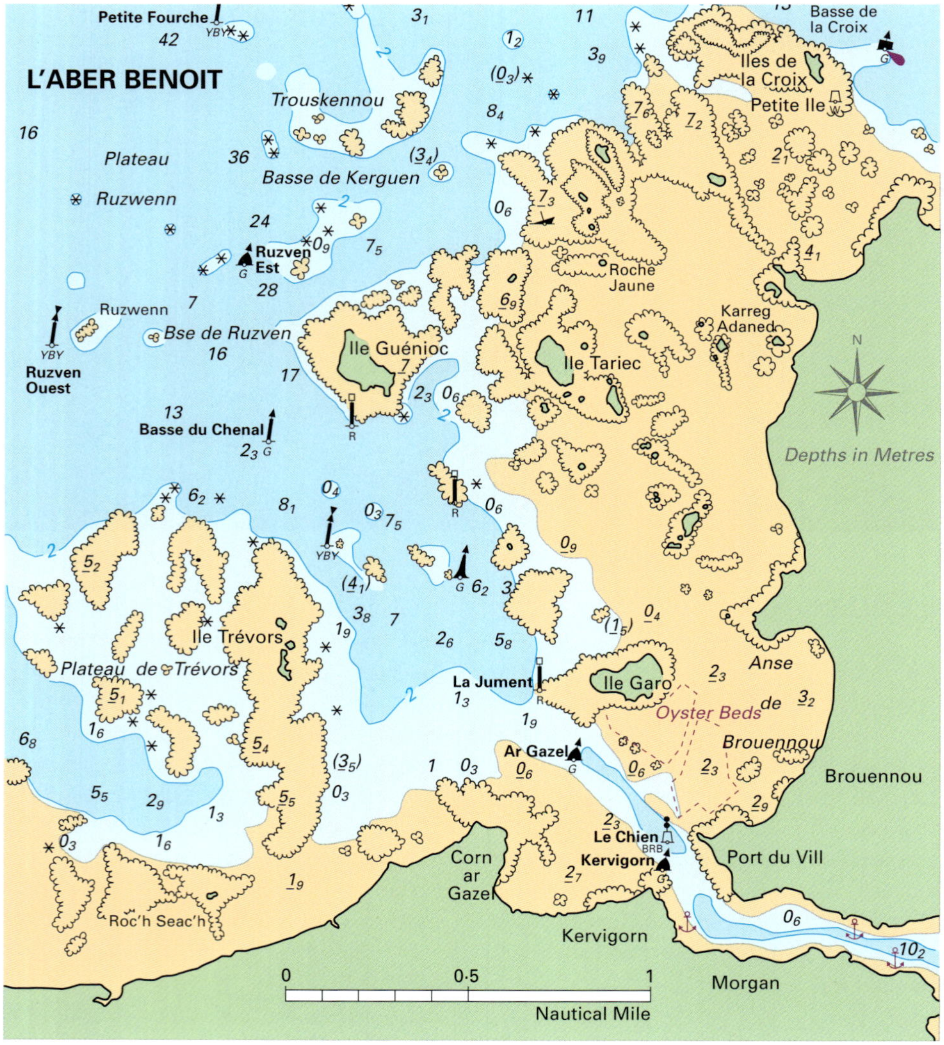

L'ABER BENOIT

This attractive unspoilt estuary lies just west of L'Aberwrac'h, barely a couple of miles SSE of Le Libenter whistle buoy, but most yachts pass it by in favour of its more spacious neighbour. L'Aber Benoît is smaller, shallower and more enclosed than L'Aberwrac'h, but worth a call if you have time to spare. Entry is possible at almost any state of tide, but above half-flood is best, and only in daylight and reasonable visibility. Avoid L'Aber Benoît in heavy swell or in strong west or northwesterly winds. From Le Libenter find La Petite Fourche W Cardinal buoy, just over half a mile to the SSW, then follow the red and green buoys and beacons south and southeast towards the narrow entrance. Once you are in the river, L'Aber Benoît offers an excellent anchorage sheltered from all quarters.

In quiet or offshore weather you can anchor off the beach, just outside the entrance on the west side. Otherwise, continue half a mile further up to Le Passage, anchoring in mid-stream clear of the local moorings, just below or just

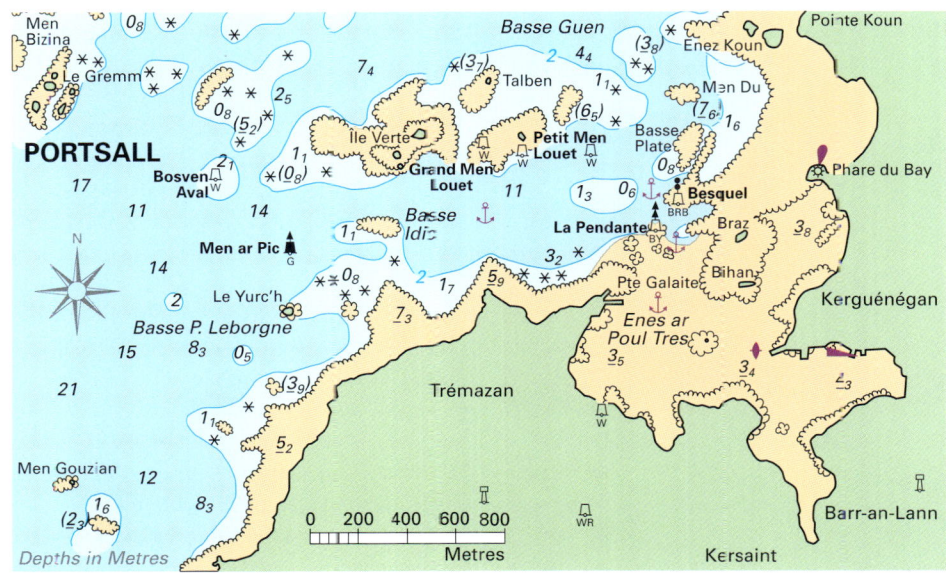

L'Aber Benoît

above Pointe du Passage and its landing slip. Don't stray more than a quarter of a mile above Le Passage, where the river starts to shoal very quickly.

PORTSALL

Most yachts tend to hurry past this rocky, rather exposed northwest corner of Brittany as they head gleefully south towards South Brittany or north,

SECRET ANCHORAGES OF BRITTANY 115

Portsall inner harbour at high water

Portsall sunset

(usually running late) at the end of a summer cruise. There are, however, various anchorages off Portsall harbour in quiet weather or with the wind from between south through southeast to east. Strangers should approach Portsall from the west via the Chenal de Men Glas, having first made a position with Le Four lighthouse bearing about 210°M distant two and a half miles. Slack water is preferable (either HW or LW Brest).

Coming from the Chenal du Four, be sure to stay well outside Roches d'Argenton, a long string of drying dangers extending two and a half miles northeast of Le Four lighthouse. Coming from L'Aberwrac'h or L'Aber Benoît, give a wide berth to Roches de Portsall, just as if you were southbound for the Chenal du Four.

Enter Portsall outer anchorage heading a shade north of east, leaving Bosven Aval rock (4.9m high) to port, Men ar Pic green beacon tower to starboard and Ile Verte south rock (4.9m high) to port. Near low water, be sure to avoid Basse Idic (awash at datum, lurking two cables east by north from Men ar Pic) and fetch up 1–1½ cables south of Grand Men Louet islet (11m high). This outer anchorage is easy to leave at night, using the white sector of Portsall light. With sufficient rise of tide you can continue just beyond Basse Karrat (0.1m depth at datum) to anchor 100 metres west of Besquel BRB beacon tower or 60 metres southeast of La Pendante N Cardinal beacon tower.

ILE DE BATZ TO L'ABER-ILDUT

Near neaps you'll find better shelter from swell by edging closer to Portsall harbour and anchoring 100 metres south of La Pendante. Use Admiralty chart No. 1432.

ARGENTON

This interesting anchorage lies one and a half miles east of Le Four lighthouse, but should only be visited in light or easterly winds in the absence of swell. From a

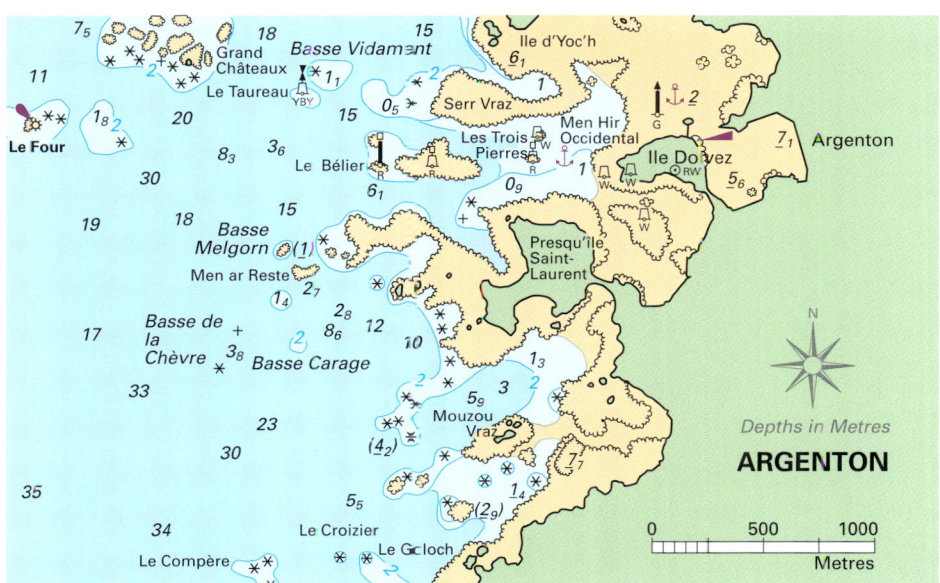

Argenton and Presqu'île Saint-Laurent

SECRET ANCHORAGES OF BRITTANY 117

position a quarter of a mile south of Le Four, make good a smidgen north of east true towards the entrance, aiming to pass midway between Le Belier red beacon, which is left to port, and Melgorn Bihan rock (10m high), left to starboard. Now pick up the leading marks – the front mark a dumpy white beacon tower in front of Ile Dolvez and the rear mark a white pyramid on the island itself. Keep these two exactly in transit bearing 083°, leaving Brividic and Les Trois Pierres red beacon towers to port. Take special care to avoid the drying rocks extending ESE from Brividic.

At ordinary springs, moderate draught yachts will stay afloat in the area between Les Trois Pierres and the front leading mark, or further north with Men Hir white beacon tower in line with Le Four. At neaps, shallow draught boats will find better shelter in the sandy bay NNE of Ile Dolvez. Use Admiralty chart 3345.

MELON

Two miles south of Argenton entrance, this small natural drying harbour is protected from the west by Ile Melon. In quiet weather, approach at half-tide from the NNW, having passed between Les Liniou and Le Four lighthouse and then kept two cables west of Le Compere rock (dries 7.2m). Le Four in transit with Le Compere astern leads into the outer anchorage a cable north of Ile Melon. Shallow-draught boats can anchor east of Ile Melon at neaps.

L'ABER-ILDUT

This sheltered natural inlet is one of the three North Brittany 'Abers' and is well covered by the pilot books, but it's worth remembering that the anchorage just outside the narrow entrance is easily accessible from the Chenal du Four at all states of tide. This attractive spot off a quiet beach offers a snug overnight berth in easterly weather, provided there's no swell rolling in from the Atlantic. The approach is relatively straightforward from the Chenal du Four, leaving Pierre de Laber green beacon to starboard, Le Lieu red beacon tower to port, and then keeping well south of the drying rocks that lie between Le Lieu and the north side of the river entrance. Anchor as near to the entrance as soundings permit and be sure to show a riding light at night. The best spot near neaps is a cable due west of the south head of the entrance, where the sandy bottom dries 1½ metres at datum. Use Admiralty chart 3345.

ILE DE BATZ TO L'ABER-ILDUT

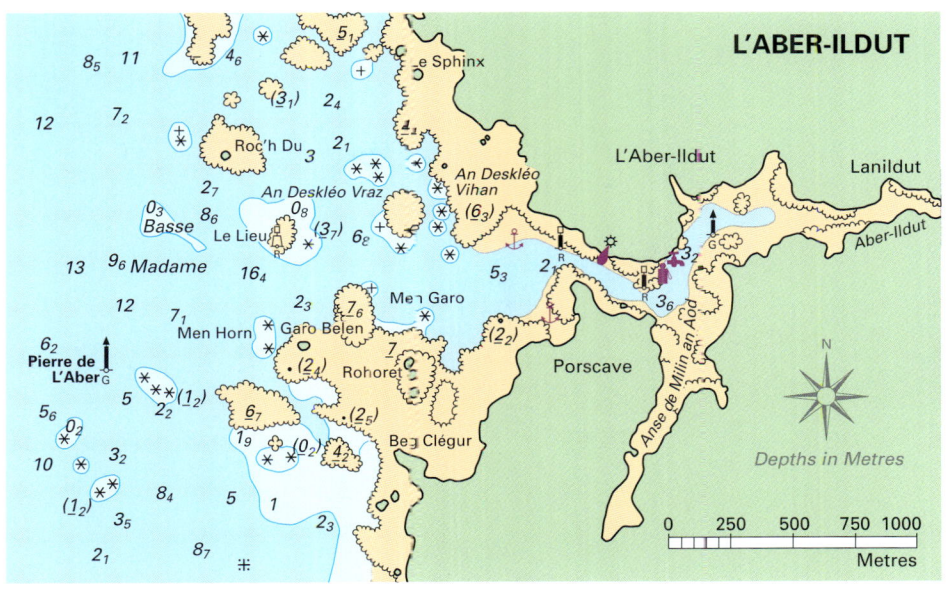

Le Lieu beacon tower

USEFUL ADMIRALTY CHARTS

No. 1432 Le Four to Ile Vierge
No. 2745 Approaches to Roscoff and Morlaix
No. 3345 Chenal du Four
No. 3668 Le Four to Anse de Kernic
No. 3669 Anse de Kernic to Les Sept Iles

USEFUL SHOM CHARTS

No. 7095 Baie de Morlaix – De l'Ile de Batz a la Pointe de Primel
No. 7150 De Portsall à l'anse de Kernic
No. 7122 De la Pointe de Saint-Mathieu au phare du Four – Chenal du Four

L'Aber-Ildut

SECRET ANCHORAGES OF BRITTANY   119

The spectacular rocky approaches to L'Aber-Ildut

CHAPTER 4

CHENAL DU FOUR TO THE RAZ DE SEIN

The wild northwest of Finistère has some of the most spectacular cruising in France. There's something almost mystical about its extravagantly indented coastline and a kind of dramatic tension hangs over these rocky waters where the far edge of Brittany faces the brooding Atlantic. Something like 100 square miles of reefs, shoals and small islands straggle seawards from Pointe de St Mathieu, ending in the gaunt profile of Ushant – a legendary landfall for seamen through the ages. The Chenal du Four cuts between the mainland and these offshore dangers, a popular route down to the charming harbours and rivers of South Brittany.

Despite its rugged character, this far-flung tip of Brittany nevertheless has more mainland shoreline that is navigationally straightforward than the north coast. Between Le Four and the Raz, there's rather more scope for edging inshore and dropping the hook than you'll find, for example, between Lézardrieux and the L'Aberwrac'h estuary. This is partly because of some long stretches of steep-to cliffs and also a good many bays that are more or less uncluttered by off-lying rocks. But you also have two natural inlets on a grand scale – the great roadstead of L'Iroise, which leads to the Rade de Brest and the Elorn and Aulne estuaries, and, further

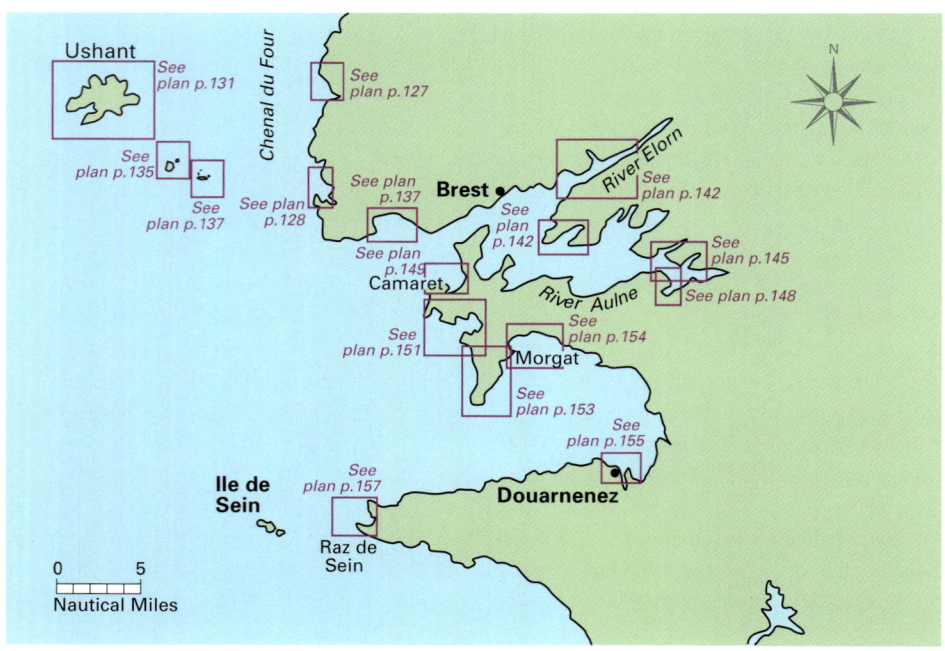

122 SECRET ANCHORAGES OF BRITTANY

south, the attractive sweep of the Bay of Douarnenez, both fascinating cruising areas in their own right.

In this chapter, starting with Portz Paul, I'll be looking at a few anchorages just off the Chenal du Four before striking offshore for Ushant, Ile de Molène and Ile de Quéménès. Then we move inland to the Rade de Brest and some peaceful river hideaways before cutting round the corner, via the Chenal de Toulinguet, into the Bay of Douarnenez. We fetch up in the Baie de Trépassés, just north of the Raz de Sein.

The Brittany islands that lie seaward of the Chenal du Four are much appreciated by yachtsmen, sometimes unknowingly, for the natural protection they give from ocean swell. Their mysterious profiles are often glimpsed out to the west as you navigate the Four, perhaps as no more than a hazy smudge on an indistinct horizon. They seem to beckon with a tempting fascination and yet, on the whole, British cruising yachts rarely visit these islands. This reticence to explore has partly to do with the somewhat forbidding atmosphere surrounding this northwest corner of Brittany, with its powerful tides and exposed position facing the Atlantic. It also has to do with timing and the fact that most yachts using the Chenal du Four are cruising to a tight schedule – either trying to get well south at the start of a Biscay cruise, or hurrying home again with time running out.

There is no doubt that, having reached the outer marks for the Chenal du Four safely, it takes a certain effort of will to direct your course out to the west, towards the islands of Ushant, Molène and Quéménès. Having done it once, though, you'll find that the solitude of the island anchorages will draw you offshore time and again when cruising past this fascinating and extremely atmospheric cruising area. This is Brittany cruising at its most satisfying and rewarding.

Le Stiff lighthouse and radar tower

Ushant – Ile d'Ouessant in French – is by far the largest of the Finistère islands, some four and a half miles long by two miles wide. As North Brittany goes, the Ushant coast is fairly steep-to, except off its southwest corner inside La Jument lighthouse, and on the north side where the Chaussée de Keller rocks straggle a good three-quarters of a mile WNW of Ile de Keller. Elsewhere, you will stay in safe water by keeping at least half a mile seaward of the nearest visible off-lying rock.

Ushant has two main harbours which are also useful anchorages: Lampaul, at the head of the long bay on the southwest side of the island; and Le Stiff, the landing place on the east side which is normally used by the mainland ferries and tourist vedettes. Both now have good visitor's moorings, but there is still room to swing to your own chain if you prefer. Le Stiff is a relatively easy point of arrival for yachts coming across from the north end of the Chenal du Four, and is reasonably protected with the wind in the west. It's a convenient anchorage to make for if visiting Ushant for the first time, and the tall radar control tower on the north side of the bay is an unmistakable landmark as you approach from the east.

However, Lampaul is rather more attractive than Le Stiff and handy for the island's shops and hotel-restaurants. It is also more sheltered than you might think from the chart since, at least in moderate weather, the long bay of Lampaul seems to filter out a good deal of any westerly swell. The sturdy visitors' moorings at the head of the bay are pretty snug in reasonably settled summer weather and there is still room to anchor further north off the lifeboat slip.

Ashore, Ushant is harsh and windswept, well dotted with houses but practically treeless. There are some dramatic coastal views, especially on the north side, and it's interesting to walk down to Créac'h lighthouse, one of the most powerful in Europe. From here, on a clear day, you can see the steady procession of ships entering and leaving the Channel via the Ushant traffic lanes.

Between Ushant and the south end of the Chenal du Four is a complex string of reefs and small, low-lying islands. Much of this area is tricky to navigate without local knowledge, but I have included two of the most accessible anchorages – off Ile de Molène and Ile de Quéménès. Molène lies about five miles seaward of the Chenal du Four, although the safest approach for strangers is from the north, leaving Plateau de la Helle to the east. The tidal streams near the island are not so savage as those around Ushant, and the waters are partly protected from swell by the surrounding reefs. Molène is only just over half a mile long, but the village on its east side is quite densely packed.

Ile de Quéménès is a couple of miles southeast of Molène, about the same length but much narrower and lying east-west. It has a few cottages, mainly towards its west end, but the east end near the anchorage and landing slip is more or less deserted. The anchorage is approached from the south and is well protected in moderate westerly weather, especially as the tide falls away and the rocky plateau to the north of Quéménès uncovers.

Back on the mainland, the sheltered Rade de Brest, the estuaries of the Elorn and the Aulne, and the higher reaches of these two rivers, offer a delightful

Baie de Lampaul

CHENAL DU FOUR TO RAZ DE SEIN

Sunrise over Molène harbour

Ile Molène

Ile Molène harbour slip

cruising ground with enough rural anchorages to keep a relaxed crew going for a month. This is an ideal area for gentle family pottering, especially when the weather at sea is cutting up rough. It also tends to be fairly uncrowded, since the detour involved in entering the Rade de Brest discourages many crews on passage, who tend to pass by outside the entrance, heading purposefully north or south. You have the facilities of the Port du Moulin Blanc marina for topping up with fuel, water and stores, and then you can lose yourself in quiet waters, tucked away in creeks and inlets.

In a similar way, the spectacular stretch of coastline between Camaret and the Raz de Sein is often taken for granted by crews hurrying past to a schedule. Yet there are several good anchorages to escape to hereabouts, either outside or inside the Bay of Douarnenez, according to wind and weather. The tides in this area are fairly weak compared with those in the Chenal du Four to the north and the Raz de Sein to the south. The pilotage is generally straightforward, especially in the Bay of Douarnenez, although you have to be a bit careful about the off-lying dangers when rounding Cap de la Chèvre.

PORTZ PAUL

This small drying harbour lies not quite two miles SSW of L'Aber-Ildut and offers an outer anchorage in quiet weather or easterlies provided there is no swell. The key mark is the islet of Grande Fourche (13m high) which lies one and a quarter-miles east of La Valbelle buoy. Approach Grande Fourche above half-tide from the direction of La Valbelle, to be sure of avoiding the Plateau des Fourches, a dangerous area of drying rocks a quarter

Le Four lighthouse in quiet weather with some breaking onshore swell

CHENAL DU FOUR TO RAZ DE SEIN

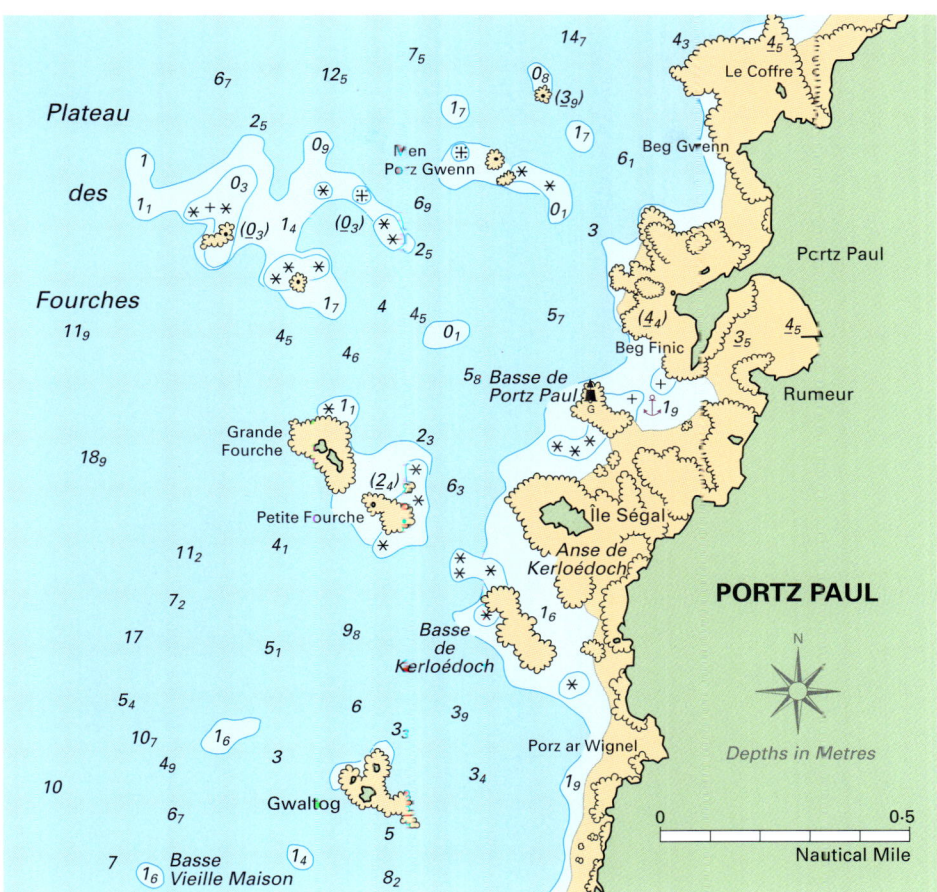

mile north of Grande Fourche. Aim to pass 1–1½ cables north of Grande Fourche (and no further), with Basse de Portz Paul green beacon tower bearing 095° dead ahead. Round this beacon tower by a cable and fetch up a cable ENE of it. Boats that can take the ground safely will be able to carry on into the harbour and dry out on firm sand clear of the local moorings.

ANSE DE PORTZMOGUER

An attractive bay off the Chenal du Four, three-quarters of a mile southeast of Pointe de Corsen. It provides good shelter from north through east to southeast and is straightforward to enter by day. When approaching from the south though, be sure to avoid Basse Jaune, a ledge of drying rocks extending nearly a quarter of a mile northwest of Pointe de Brenterc'h. There are also numerous crab-pot floats in the area. Portzmoguer has a good many small boat moorings and it's usually best to anchor outside them, close under the low headland on the northwest side of the bay. This anchorage is easy to leave at night, provided you keep a sharp lookout for pot floats.

PORTZ ILLIEN

A small bay half a mile SSE of Pointe de Brenterc'h, on the northeast side of Anse des Blancs Sablons. Illien is sheltered from north through east to southeast,

SECRET ANCHORAGES OF BRITTANY   127

La Grande Vinotière red tower off Le Conquet

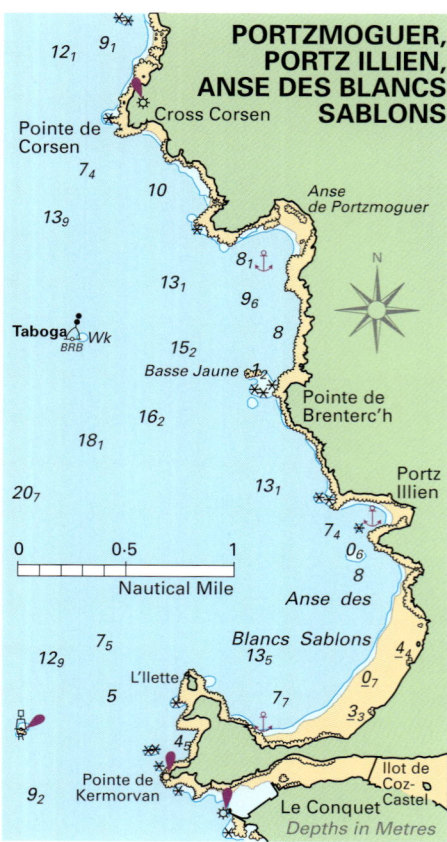

L'Ilette when the ebb is setting strongly onto it. Use Admiralty chart No. 3345.

although it is not quite so easy to enter as Portzmoguer. An isolated rock (drying 2·7m) lies just over a cable northwest of the south head of the entrance and a ledge of drying rocks extends over a cable west and 100 metres south of the north head. The easiest time to enter is near low water, when you can leave the rocks off the north side of the entrance close to port and be sure of leaving the isolated rock well to starboard.

ANSE DES BLANCS SABLONS

A wide sandy bay on the north side of Presqu'île de Kermorvan. There's a useful anchorage in the southwest corner of Anse des Blancs Sablons, sheltered from southwest through south to southeast. Approach is straightforward, but be sure to give a wide berth to

USHANT

This rather forbidding but intriguingly atmospheric island is best visited at neaps in quiet weather and good visibility. On the chart Ushant looks like a truncated lobster, its two claws pointing southwest to enclose the long Baie de Lampaul. At the head of this inlet, the close-knit town of Lampaul is the focus for island life and a small drying harbour shelters a few local fishing boats. The east side of Ushant, the lobster's tail, has its back to the prevailing weather, so the mainland ferries usually land at the solid stone jetty at the head of the Baie du Stiff.

When you first see Ushant from offshore, there seems little evidence of habitation of any kind. The sombre east cliffs look forbidding from a distance, and even on a warm sunny day it takes

Anse des Blancs Sablons

some will-power not to turn back towards the Brittany mainland. But as you get closer and peer through the glasses, small stone cottages come into focus, hinting at the possibility of friendly life ashore. At the lower west end of the island, around Lampaul, the high-pitched roofs are more numerous, clustered in groups for moral support against the elements. Because of its excess of bracing sea air there are virtually no trees on Ushant, but a few shrubs survive in sheltered corners.

The Baie de Lampaul feels rather like a sound on the west coast of Ireland. Up on the wild shore of each enclosing arm, crofts are dotted here and there among the scrub and tufty grass. Off the end of each headland stands a gaunt lighthouse, built heaven knows how in these swirling currents. La Jument stands a mile southwest of the southern headland, Nividic lighthouse not quite half a mile off the northern spur, Pointe de Pern.

At the head of the bay, the ornate spire of Lampaul church has a strangely old-fashioned profile and the lifeboat house testifies to the power of the sea. Yet on a still summer day, you'll be glad you persisted with this landfall. To have arrived at this extraordinary island with your own boat is one of the special rewards of Brittany cruising.

Lampaul There are good visitors' moorings right at the head of the long inlet of Lampaul, and various anchorages around the Baie de Lampaul depending on the wind direction. My favourite spot is still right in the northeast corner of the bay, fairly close southwest of the lifeboat slip. Within a couple of hours of high water, it's easiest to land in the tiny inner harbour, home to a few local fishing boats and drying

CREAC'H LIGHTHOUSE

The windswept island of Ushant – Ile d'Ouessant in French – lies 10 miles off the northwest tip of Brittany, scoured by powerful tides and holding the front line against Atlantic weather. Mariners through the ages have known the name and often seen the light, since Ushant is the classic landfall and departure point for the south side of the English Channel. In the days of sail, almost all outward-bound shipping had to clear Ushant before heading south in search of favourable trade winds.

The present traffic separation lanes pass five miles seaward of Ushant, so the island's hard profile is still glimpsed by merchant crews every hour and day of the year. Yachtsmen often spot a hazy silhouette as they hurry through the Chenal du Four between the Channel and Biscay. Indeed Ushant seems curiously familiar to anyone who goes to sea, yet few outsiders ever land there, or even think of the place as a real island with a village of inhabitants all year round. Every night, summer and winter, this remote community turns in beneath the piercing sweep of Créac'h light, the most powerful navigation beacon in Europe.

Creac'h lighthouse

This commanding black and white tower is arguably the most dramatic lighthouse around the French coast. It stands on the northwest edge of Ushant, where Atlantic gales howl across the cliffs at fever pitch. The old generator rooms in the imposing lighthouse building have been converted into a fascinating museum – the Centre d'interpretation des Phares et Balises.

The French do these things well. As you wander round the cavernous rooms, the life-and-death significance of lighthouses through the ages becomes immediately clear, sharpened by the fine attention to detail and year-round routine that is so apparent from the exhibits. To study the magnificent engineering of a vast rotating light is a true aesthetic experience. You can see the massive three-cylinder diesel engine that once powered the lighthouse generators and also supplied the whole island with electricity until underwater cables from the mainland arrived in 1988.

CHENAL DU FOUR TO RAZ DE SEIN

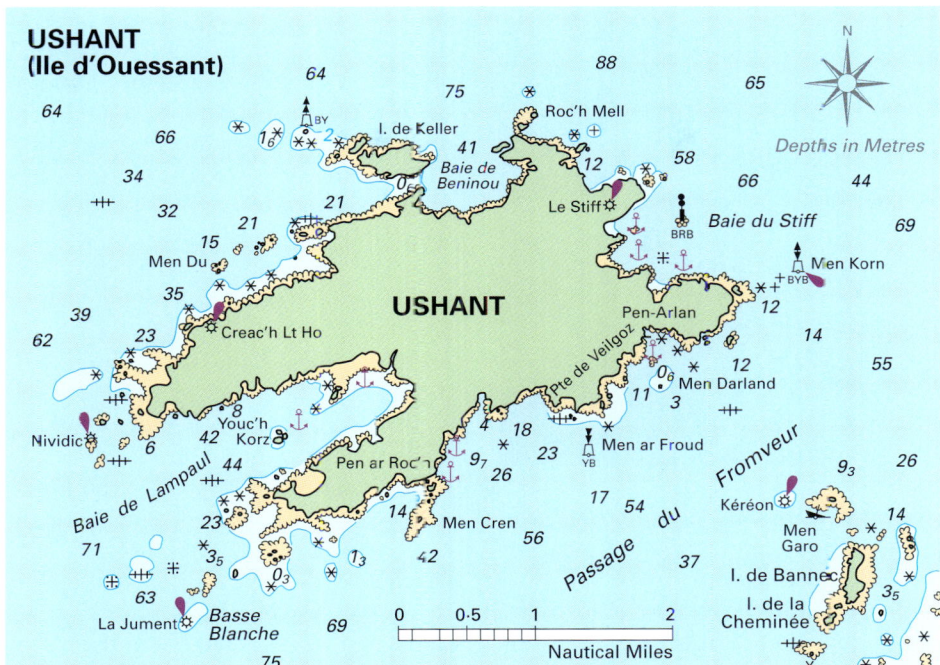

Baie de Lampaul

SECRET ANCHORAGES OF BRITTANY   131

Lampaul inner harbour on the southwest side of Ushant

Baie de Lampaul

soon after half ebb. Near low tide, you can take the dinghy to the outer jetty just west of the lifeboat slip.

Baie du Stiff On the east coast of the island, the Baie du Stiff also has a few moorings and offers reasonable shelter in westerlies, with the advantage of making a fairly straightforward landfall direct from the north end of the Chenal du Four. It's usually best to tuck into one of the three smaller inlets within the Baie du Stiff – Portz Liboudou, Portz an Dour or Portz Aheac'h – so that you are clear of the mainland ferry coming and going, which it does with panache.

Portz Darland Tucked into the southeast corner of Ushant, Portz Darland is well protected in northwesterlies. Approach from the SSE, preferably near low water when the various drying rocks on either hand are well exposed. From at least half a mile offshore, identify the stone jetty on the west side of Portz Darland and the sandy beach just east of the jetty. Bring Le Stiff lighthouse into line with the east end of the beach bearing 340° and close the shore on this transit leaving Men Darland rock (drying 7.5m) and Fret Kas rocks (drying 5.8m) each a cable to

CHENAL DU FOUR TO RAZ DE SEIN

Le Stiff radar tower

Le Stiff lighthouse

port. Anchor off the jetty, as close to the beach as your draught allows.

Baie de Pen ar Roc'h About one and a half miles WSW of Portz Darland, the rugged looking Baie de Pen ar Roc'h has anchorages on its west side about a cable north of Roc'h Nel (10m high) or further north in the Anse de Boug an Dour. Approach from southeastward near slack water, leaving Roc'h Nel a cable to port. French SHOM Chart No. 7123 is invaluable for approaching and exploring Ushant.

ILE MOLÈNE

This little-visited island lies between Ushant and Le Conquet in the midst of what seems like a rather formidable area of drying rocks and powerful tides. However, in quiet weather, the snug harbour at Molène is easily approachable from due north, by making for Les Trois Pierres lighthouse first and then leaving it a cable to the east to pass between Bazou Réal E Cardinal beacon tower and Roche Goulin W Cardinal buoy.

There is also an anchorage about half a mile SSE of Les Trois Pierres, close to the east of the string of rocks and islets known as Les Lédénez de Molène. The safest approach is again from the north, making for Les Trois Pierres at first but then passing half a mile east of the lighthouse and Petit Lédénez de Molène. Fetch up in the natural bight between Grand and Petit Lédénez, about two cables from each. This is a delightful spot in gentle westerly weather. Use French SHOM Chart No. 7122 and aim to arrive near slack water if possible.

Local boats dried out at Ile Molène

SECRET ANCHORAGES OF BRITTANY

THE WRECK OF THE *DRUMMOND CASTLE*

The northwest Brittany islands of Ushant and Molène are wild and windswept, surrounded by countless reefs and scoured by the powerful tides that pour round this corner of France to and from the English Channel. Atlantic gales often make their landfalls here and a long Atlantic swell can give these waters an eerie feel even in calm weather. This far tip of Brittany is also notorious for fogs.

One of the most chilling sea stories of these islands is the wreck of the Castle Line sailing packet *Drummond Castle* in 1896. This sizeable vessel was carrying 246 souls between South Africa and England when, at 2230 on 16 June, she struck the Chausée des Pierres Vertes reef about four miles south of Ushant. The *Drummond Castle* went down in only a few minutes. Only one passenger and two of the ship's company survived. Two particular islanders played a crucial role in saving these three – a retired Ushant fisherman, Joseph Berthelé, and Mathieu Masson, skipper of the Molène fishing boat *Couronne de Marie*.

When the *Drummond Castle* struck Les Pierres Vertes the weather was quiet but the islands were shrouded in thick fog. Most of the passengers were English women and children evacuated from South Africa at the start of the Boer War, so the tragedy of this wreck was especially poignant. There was no radio communication in 1896 and no flares could be seen in the fog. In any case, the ship foundered very quickly, before any kind of emergency action could be taken.

Because the fog was still dense next morning, no fishing boats went out from either Molène or Ushant. It was only when someone on Molène chanced to hear a plaintive cry for help drifting through the mist that Mathieu Masson set off in his boat to investigate, navigating partly by instinct and partly by dead-reckoning using a clock and a compass. Following the cries, by now growing fainter, Masson eventually came across an officer and a sailor clinging to an isolated rock. These men were in poor shape and Masson took them carefully back to Molène. It wasn't until the officer had recovered a little and was able to speak that the islanders realised what had happened and that 244 other passengers and crew were still out there somewhere, in or on the cold clammy sea. They organised a concerted search and sent someone by boat to alert the locals on Ushant. Later Joseph Berthelé found a solitary surviving passenger clinging to a piece of wreckage, but nobody else was saved.

Between 1880 and 1920 there were 171 wrecks off these islands, most of which were English ships. Ushant has always been a crucial and tricky navigational corner for shipping, set as it is just at the entrance to the English Channel, and mariners through the ages have tried to pass this gaunt island close enough to identify it but not so close as to be dangerous.

Today's navigators may be equipped with every possible high-tech aid and they are watched by the radar-tracking station on Ushant, but mariners still keep a wary eye on Ushant. However minuscule the risk of shipwreck these days the possibility is still always there.

ILE MOLÈNE

With careful pilotage you can approach Molène from the east more directly from the lower reaches of the Chenal du Four. From a position half a mile northeast of Pourceaux N Cardinal buoy, pick up the leading line that guides you through Chenal des Laz towards Molène. This is a classic Brittany transit – a white post on an islet in line with a ruined mill bearing 264° – and you should pick up these marks even when approaching using GPS waypoints. Just over a mile from Molène, you can head just north of west towards the Lédénez outer anchorage, or turn north to skirt the islets and reefs in front of the island before approaching the harbour from the NNE.

The closely-packed cottages around Molène harbour look very Breton with their pitched roofs and shuttered windows, but the faded stone and weathered paint also show that this remote village is in the front line of Atlantic weather. The outer jetty you pass to enter the harbour is where the ferries come in from Brest and Le Conquêt. To the south, a shorter but older stone pier protects the shallow inner harbour. You'll often see nets

Ashore at Ile Molène

spread along the quay, and behind them is the whitewashed frontage of the island's only hotel, Kastell an Daol.

ILE DE QUEMENES

A small low-lying island two miles southeast of Molène. There is an attractive anchorage in the Passe du Cromic between the east end of Quéménès and Ile de Lytiry. Approach from the southwest, having reached Pierres Noires S Cardinal buoy three-quarters of an hour before HW Brest and then made good 040° for five miles to the entrance to Passe de Cromic. This track leaves La Vieille Noire E Cardinal beacon half a mile to port. Enter Passe de Cromic from due south and fetch up about 100 metres northeast of the Quéménès slip in a metre depth LAT. Although this is a spot for settled weather and good visibility, there is shelter from the W once you are inside. Use French SHOM Chart No. 7122.

ANSE DE BERTHEAUME

This broad bay lies just over three miles east of Pointe de St Mathieu and the anchorage in its southwest corner is well sheltered in westerlies and north-westerlies. Fetch up outside the local moorings and be sure to avoid Le Chat, a nasty reef (dries 6.8m) lurking a cable northeast of Fort de Bertheaume. You can land at the slip on the south side of the beach and 20 minutes' stroll takes you to the village of Plougonvelin. Admiralty chart No. 3427 is useful.

CHENAL DU FOUR TO RAZ DE SEIN

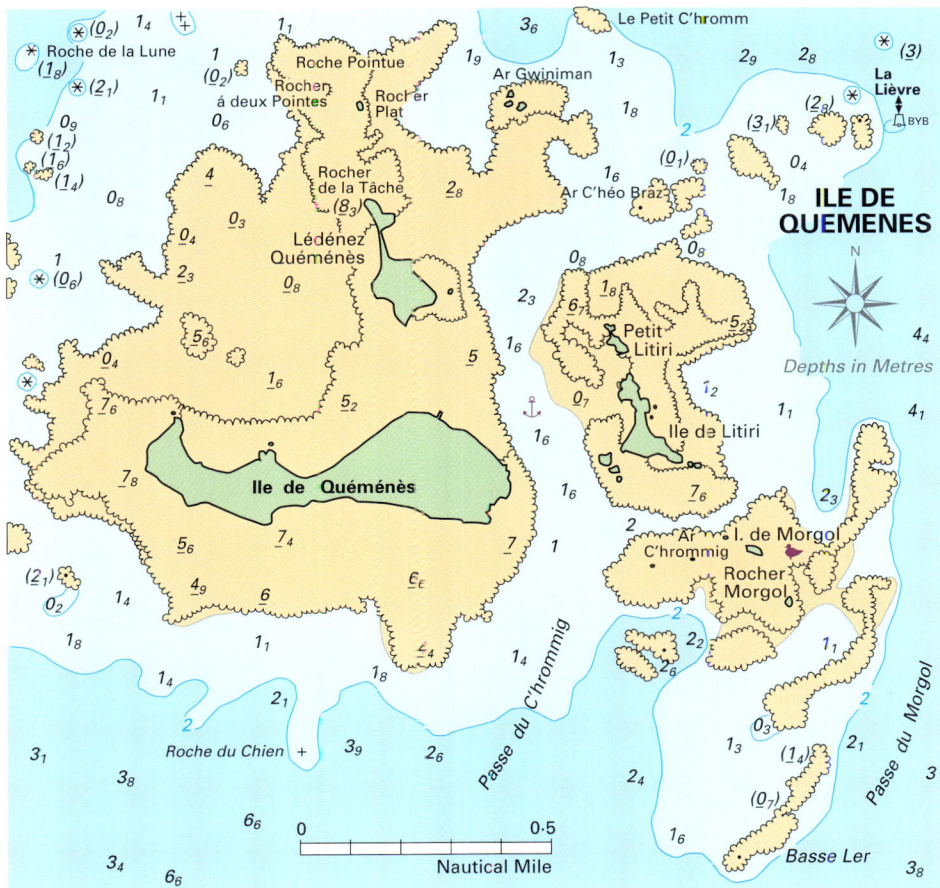

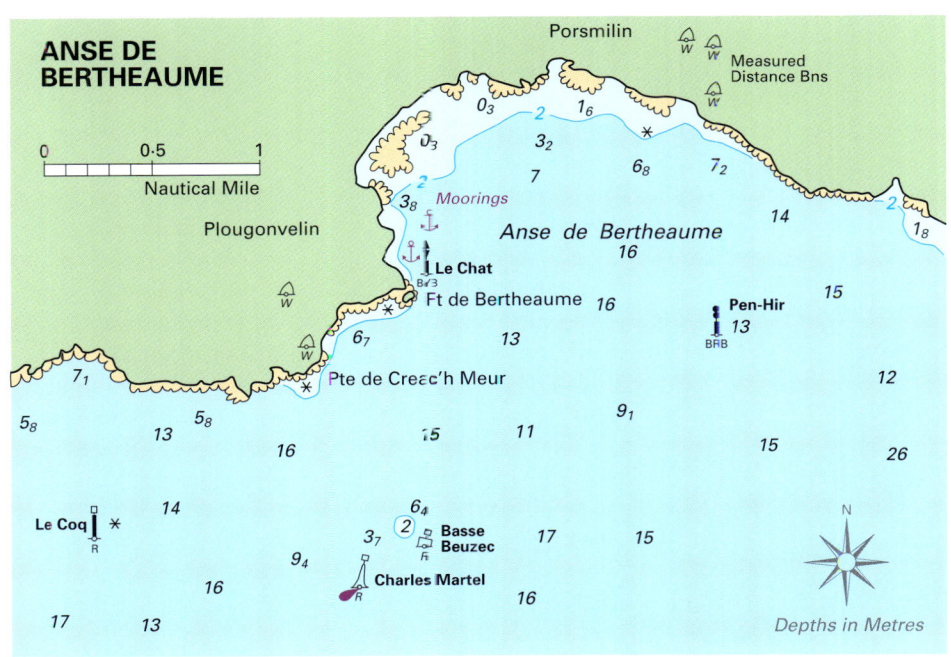

SECRET ANCHORAGES OF BRITTANY 137

CRABS OF MOLÈNE

Fresh crab is probably my favourite seafood, which I hope would be plentiful if I were ever cast away on a desert island. Crab meat has a succulence and depth of flavour far superior to that of its costly colleague, the haughty lobster. The heady zest of the sea in a crab evokes cruising memories at the very first taste, transporting you to tidal waters and dramatic rocky coasts even if you are picking the shell expensively in a city restaurant.

We enjoy superb Devon crabs at home, but some of the finest crabs I've ever eaten while cruising have been caught in the tide-swept waters around Ile Molène. This tiny French outpost lies seven miles off the northwest corner of Brittany, one of a string of islands and reefs that help shelter the Chenal du Four from the Atlantic. UK boats heading for South Brittany pass within five miles of Molène as they run down the Four towards Brest, although navigators are usually too preoccupied with beacons and waypoints to pay much attention to this low and surprisingly populated 230 acres of close-knit community lurking to seaward. But if you enjoy attacking a good crab, make for the restaurant of the only hotel on Molène – the Hôtel Kastell an Daol – run with traditional expertise by the Masson family.

Hen (female) crabs are by far the tastiest in my view, and in their prime have three different types of meat within the same shell, providing an amazing range of flavours. The fleshy white chunks in the two main claws make generous mouthfuls. Tucked well inside the body shell, the brown creamy meat is a luxurious vein of pleasure, with the smoothness of rich pâté and the salty tang of a low-tide estuary. Hen crabs in the autumn have mouth-watering red coral meat alongside the cream, slightly crumbly parcels of eggs with a hint of caviar but firmer, as if red caviar had been pressed into small steaks, lightly grilled and served cold.

A good French mayonnaise goes well with crab, but a slightly sharp vinaigrette perfectly complements the richer meat. To make this on board, finely chop two shallots, put them in a small bowl and add half a cup of red wine vinegar. Stir in half a teaspoon of paprika, season with salt and pepper and let the mixture stand for about half-an-hour before using. A little of this liquor spooned over some freshly picked crab meat makes a sublime combination.

I always choose a Gros Plant du Pays Nantais with crab, a crisp Loire wine produced from the Folle Blanche grape of the Muscadet area. Gros Plant is sharper than a Muscadet and often a limpid green in colour. It's an inexpensive wine, but don't eschew it on menus simply on that account. Gros Plant should be served very cold and its clean flavour is an ideal foil to rich crab meat.

Pointe de St Mathieu

Fort de Bertheaume

LES MOTTES D'OUESSANT

Ushant's traditional stone cottages, their tiny windows protected by stalwart shutters, seem literally to huddle against the unfettered winds that accelerate over the Ushant cliffs and cut across the treeless landscape. If you are walking out to Créac'h lighthouse from Lampaul harbour, it's worth calling at the small museum in the Maison du Niou, where the austere living quarters of a traditional Ushant cottage are splendidly preserved. Inside the two simple rooms, you can see the original dressers and cupboards made from driftwood recovered from the sea. Ushant has practically no trees, so floating or beached timber was a valuable commodity. Most of the furniture is painted in the traditional Ushant blue and white. The cottage floor is hard-trodden earth and the partially enclosed beds look like fishing boat bunks, designed to keep occupants warm and dry in hostile conditions. La Maison du Niou is about 20 minutes' walk from Lampaul and is open from mid-morning during the season.

Because Ushant has no trees, and hence no firewood, the islanders traditionally burned rough turfs – *les mottes* – for cooking and heating, in much the same way as Irish crofters. The *mottes* were cut from around Ushant's cliffs and dried in summer for use in the winter. Most Ushant men went away to sea, leaving their wives to grow vegetables on tiny smallholdings and graze sheep on the scant grass. Mutton and vegetables were the staple foods, made into a *ragoût* cooked long and slow in simple *motte* ovens. The Ushant women would put this stew together in the mornings, leaving the heavy iron casseroles to simmer while they were out in the fields.

This succulent Ushant dish is still produced to great effect at the Hôtel Roc'h ar Mor in Lampaul, where they still cook *ragoût d'agneau* in a *motte* oven just outside the hotel. Not to be missed.

CHENAL DU FOUR TO RAZ DE SEIN

THE BREST BLOCKADES

The Rade de Brest is one of the great roadsteads of naval history. During the Seven Years and Napoleonic Wars, the sheltered reaches inside Presqu'île de Quélern were packed with French ships waiting for the right conditions to escape the blockade of British squadrons that patrolled continuously for years on end. Now, as you round Pointe de St Mathieu and turn in towards Brest, a few 18th and 19th-century spirits still hang in the air. The coast looks wild and the abbey ruins up on the point seem to roll the landscape back in time. Training the glasses through the narrow Goulet de Brest, you'd not be surprised to glimpse the crossed yards of anchored 74s or the graceful lines of an East India trader. Fans of C S Forester's *Horatio Hornblower* will be able to picture the scene in 1803, when Admiral Cornwallis began his gruelling blockade of Brest with a detachment of frigates and ships of the line from the Channel Fleet. The 1803-1815 Napoleonic Wars were a tremendous period for British naval expansion and Britain's supremacy at sea was aptly symbolised by the Brest blockades, which prevented French warships from getting into action and strangled supply lines to what was then France's principal naval port.

Cruising this area now, you can still see clues from that desperate era. On the north shore, Fort de Bertheaume once guarded a strategic anchorage where French coasters might huddle if they'd managed to sneak into L'Iroise at night. Three miles further east, Petit Minou lighthouse marks the mouth of the Goulet de Brest, a narrow but vital gap that was powerfully protected by forts and batteries on both shores. Even with a fair wind, sailing ships couldn't penetrate this crucial entrance without being caught by the crossfire of heavy guns.

Some of the most fascinating aspects of the Brest blockades in the days of sail turn on the seamanship involved in keeping a fleet on station all year round in all weathers. Although the Brest approaches have some stretches of easily navigable water, this is a rocky corner of Brittany. North of L'Iroise is the tangle of islands and reefs between St Mathieu and Ushant. To the south, the dangerous Chaussée de Sein reaches out from Pointe du Raz for 13 miles. Further inshore there are plenty of shoals to threaten sailing ships that might suddenly find themselves hemmed in by a wind shift or helplessly becalmed.

In easterly winds, when the French would be most likely to try and break out, the British could also afford to work close inshore to prevent this. In strong westerlies, the French couldn't hope to escape through the narrow Goulet, but the blockading fleet had to beat well *offshore* to gain precious sea room, especially during winter gales. The trickiest periods for the British occurred when conditions were just starting to improve after a spell of heavy weather. A sudden veer might give the French ships just enough slant to sail out through the Goulet while the British were still at offshore stations.

Old Brest fort

Le Petit Minou lighthouse

SECRET ANCHORAGES OF BRITTANY

ELORN RIVER

This rather pleasant river flows into the Rade de Brest just east of Moulin Blanc marina. The Elorn is navigable at any state of tide for two or three miles above the two high road bridges – Pont Albert Louppe and Pont de l'Iroise – to just beyond the moorings off Anse St-Nicolas. There are various possible anchorages in these lower reaches, which few cruising yachts visit. Use Admiralty chart No. 3429.

Le Passage This attractive spot, once a ferry crossing point, lies about three-quarters of a mile above the bridge on the south side of the river. You should fetch up clear of the local moorings and buoy the anchor, or you can sometimes use a vacant mooring if it seems strong enough.

Anse St Nicolas About a mile further upstream from Le Passage, you can anchor off the north shore of the Elorn in the Anse St Nicolas, which offers better shelter than Le Passage if the wind is at all fresh from the southwest.

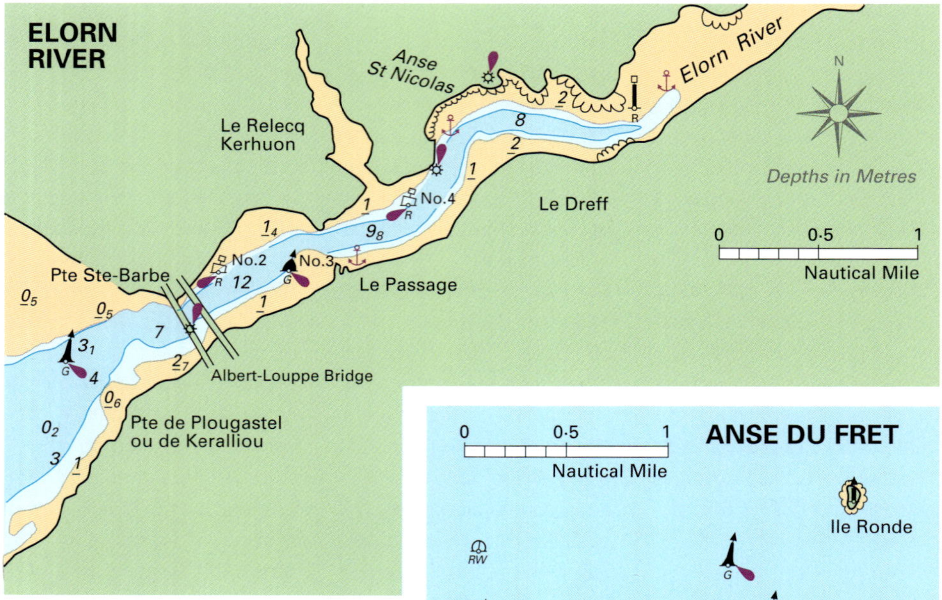

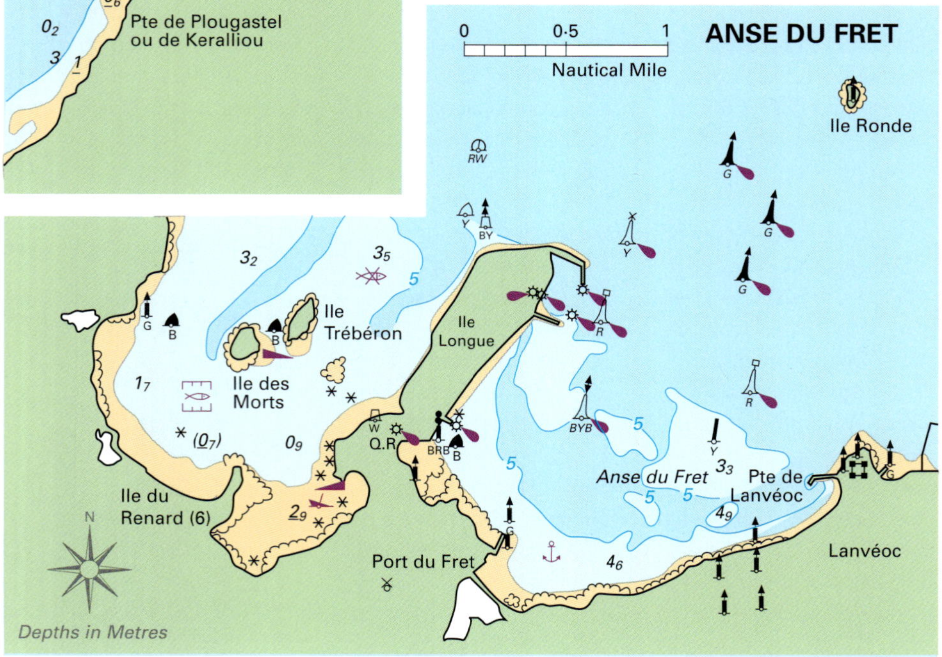

The Elorn river bridges

Kermeur St Yves A mile above St Nicolas you can find good shelter from all quarters around neaps in the last tongue of deepish water, about a quarter of a mile northeast of St Jean red beacon tower.

Landerneau On a reasonable tide you can get right up to Landerneau, a mellow old Breton town four miles above Anse St Nicolas. The lifting bridge half a mile below the town opens on request if you phone ahead (Port de Landerneau ☏ 06 11 03 31 20). You can moor at the pontoon below Pont Caernafon on Quai Barthélémy Kerros, but don't linger long after high water if you can't take the ground safely – the river dries to a trickle. If you can dry out, the best spot is against the wall on the north shore just below the fixed road bridge that determines the head of navigation.

Further up in the town is a rather fine 16th-century bridge with shops built on it, and an old commercial quarter from the 17th and 18th centuries. There are plenty of shops, cafés and restaurants.

Drying quay up at Landerneau

AULNE RIVER

The Aulne winds inland south and east of Brest, offering some delightful anchorages in the lower reaches and navigable for almost 12 miles up to the old market town of Châteaulin. The protected waters of the Aulne are one of the great attractions of the Rade de Brest, with enough scope for a week of relaxed cruising if the weather is hectic outside.

Anchored in the Aulne opposite a sleepy Brittany village – Trégarvan perhaps, or Le Passage well inland – you feel right in the heart of rural France and time slips back naturally to a more civilised pace. Even with gale warnings out for Biscay, you are perfectly snug from the rigours of the sea. On a warm summer day, the slow pulse of the river is a soothing antidote to the babble of the 21st century. You can land at an old stone slip and wander ashore to find a village baker, a *charcuterie* for some local pâtés and a general store for wine and cheese. If you stop listening to the radio and switch off the mobile phone, the gentle passage of the days is marked only by the time between meals, the turn of the tide and the distant chimes of a village clock.

In the lower Aulne, much of the south shore of the Rade de Brest is occupied by the French military, at least between Roscanvel and the naval college opposite Anse du Poulmic. However, you can safely enter Anse du Fret and anchor opposite the attractive, typically Breton quays of Port du Fret.

ANSE DU FRET

Enter Anse du Fret from the north-east, keeping fairly close to the Lanvéoc shore and well away from the Ile Longue side of the bay, which is an important French military base. Depending on soundings, anchor east or south-east of Port du Fret clear of the local moorings. You can land

Aulne River

at one of the quays and the pleasant village has small shops and a couple of restaurants.

TINDUFF

A couple of miles further upstream from Auberlac'h, the north side of the estuary falls away into the Baie de Daoulas, where I have often anchored overnight. Much of this bay is shallow, but you can pick out the deeper patches using Admiralty chart No. 3429. As you come into the bay, avoid the shoals and a drying rock off the first headland, Pen à Lan, marked by a spar beacon. On the west shore, the commune of Tinduff has some moorings and a landing jetty. You can stay afloat here in quite a wide pool, but it's safest to approach above half-tide. At neaps, yachts of moderate draught will be able to tuck further in past the jetty and anchor clear of the local moorings, where you'll find good protection from between southwest through west to NNE.

DAOULAS RIVER

The mouth of the Daoulas River opens off the east side of the bay across a sandbar. Most of this peaceful river dries to a narrow trickle, but you can just stay afloat at dead neaps close off the south shore less than a mile from the mouth.

Enter the Daoulas River above half-tide, preferably on the flood, and anchor not quite three-quarters of a mile above Pointe du Château close off the south bank near a small quay. This rural hideaway feels miles from anywhere and offers perfect shelter from all quarters in about 1½m LAT. You'll find a sleepy café up the hill at Gorrequer and half-an-hour's walk takes you to the small town of Logonna-Daoulas.

STYVEL

About three miles above the Baie de Daoulas, you pass the steep promontory of Landévennec to starboard, with its abbey ruins on the hillside. Then the river makes a tight curve back to the

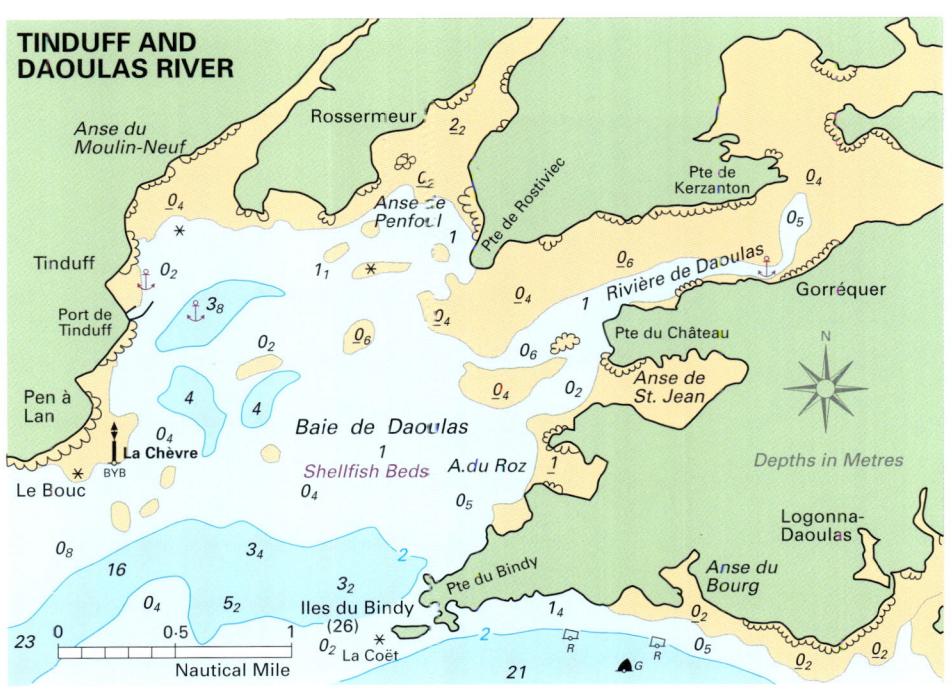

DRIFTING UP THE AULNE

As you follow the Rade de Brest south and east towards the Aulne River, you'll see an ancient abbey up on the south shore at Landévennec, as the channel starts to narrow beyond Daoulas. This rather splendid retreat was founded by a Benedictine monk, St Guénolé, in the 5th century, destroyed by the Normans in 913 and later rebuilt and enlarged. Under the protection of the Counts of Cornouaille the abbey flourished until its decline in the 18th century. There is still a Benedictine monastery here, but since 1990 there has been a museum on the site of the old monastery. To visit the abbey, you can anchor with sufficient rise of tide off the north side of the Landévennec promontory, inside No. 7 green buoy opposite the drying jetty at Port Maria. Landévennec is an attractive village with a *crêperie* in the Hotel Le Beausejour, the restaurant Le St Patrick and a small shop. Round on the south side of Landévennec, the river curves south and southeast between steep wooded banks, under the high Térénez suspension bridge. Just below the bridge there are various possible anchorages off the west bank, tucked out of the main run of tide. This is a good spot to spend a quiet night if you plan to continue upriver to the locked basin at Châteaulin – there's a good restaurant on the east bank and you'll be well placed to time the tide up to the lock at Guily Glas. The lock operates for two and a half hours each side of HW Brest between 0600 and 2200 local time. To carry the flood nicely up to Guily Glas, you should be passing under Térénez bridge about three hours after low water Brest. From the bridge to the lock is about ten nautical miles and the trip must be taken slowly.

Cruising gently up the middle reaches of the Aulne, past Trégarvan and Le Passage, you find that hardly a sign of the 21st century has penetrated to the riverside. In Napoleonic times, coasting barges would have ghosted through this same landscape under sail, transporting bulk supplies well inland with consummate ease.

Further up you pass a busy trunk road before the river falls peaceful again on the narrow stretch through reeds and marshes to Guily Glas. The lock is on the north side of the weir, just downstream from an imposing railway viaduct. There's a small waiting quay just outside the gates. The lock itself is always well kept, with picturesque flower borders and hanging baskets. You emerge into the canalised river about a mile below Châteaulin.

Moored alongside at Châteaulin

The sheltered quays at Port Launay

Daoulas River

Landévennec

CHENAL DU FOUR TO RAZ DE SEIN

The Aulne River and the abbey at Landévennec

Guily Glas lock and the impressive viaduct

west under the south shore of Landévennec. Some quite large ships are laid up in this deep, sheltered reach, rusting away in what has become a local graveyard, waiting for a place at the breaker's yard. Styvel is the first sheltered anchorage in the Aulne river proper, just opposite Ile de Térénez and

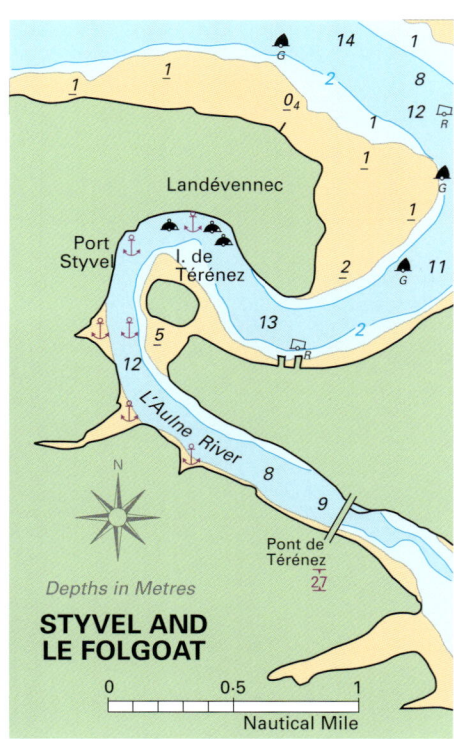

148 ⚓ SECRET ANCHORAGES OF BRITTANY

very close off the N bank. There is deep water here at all states of tide, but it's important not to fetch up too near the various large mooring buoys. Your anchor should be buoyed, in case you are unlucky enough to foul one of the ground chains.

LE FOLGOAT

There are various anchorages between Styvel and the suspension bridge, but the bottom is rocky away from the banks. There is good holding in mud at the mouth of either of the two drying inlets on the west side of the river, tucking in as close as the tide allows. A slight bay, just over half a mile below the bridge on the west side, is also a good spot if you edge well inshore. On the east side of the river, you can anchor just downstream of the local moorings that are laid just above Ile de Térénez. There are several anchorages in the Aulne above the suspension bridge, with landing places at Tregarvan and Le Passage.

ANSE DE CAMARET

Back down on the coast, tucked behind Pointe du Grand Gouin on the south side of the Brest approaches, Camaret is a natural staging post between open sea and the sheltered inner reaches of the Rade. The attractive marinas at Camaret are popular with yachts bound between the English Channel and the summer cruising grounds of Biscay, but those who prefer some seclusion overnight can anchor at various spots around the Anse

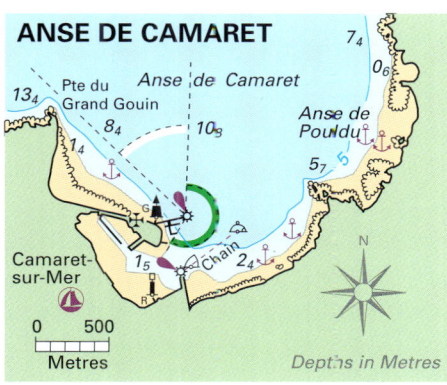

Anse de Camaret

The distinctive red stone tower of Fort Vauban at Camaret

de Camaret, depending on the wind direction. In calm summer weather or with the wind in the south, it can be pleasant lying off the sweep of beach just north of the main Camaret breakwater, where the swimming is excellent over the gradually shelving sandy bottom.

In easterlies or southeasterlies, I've often anchored for the night off the southeast shore of the Anse de Camaret, about a quarter of a mile southwest of Pointe St Barbe and as close inshore as the tide allows. Another good spot in easterlies is the Anse du Pouldu, a quarter of a mile northeast of Pointe St Barbe.

ANSE DE PEN-HIR

An attractive anchorage in a sandy bay just east of Pointe de Pen-Hir. Entry is straightforward from the Chenal du Toulinguet, by rounding Les Tas de Pois, a conspicuous group of five islets straggling for half a mile southwest of Pointe de Pen-Hir. The approach from the direction of the Raz de Sein is also straightforward if you make for Les Tas de Pois first. If coming from the south round Cap de la Chèvre, it's important to avoid Le Bouc, a low above-water rock marked on its west side by a west-cardinal buoy, and Le Chevreau (dries 6.1m) marked by a W Cardinal spar beacon.

The glorious beach at Anse de Pen-Hir

CHENAL DU FOUR TO RAZ DE SEIN

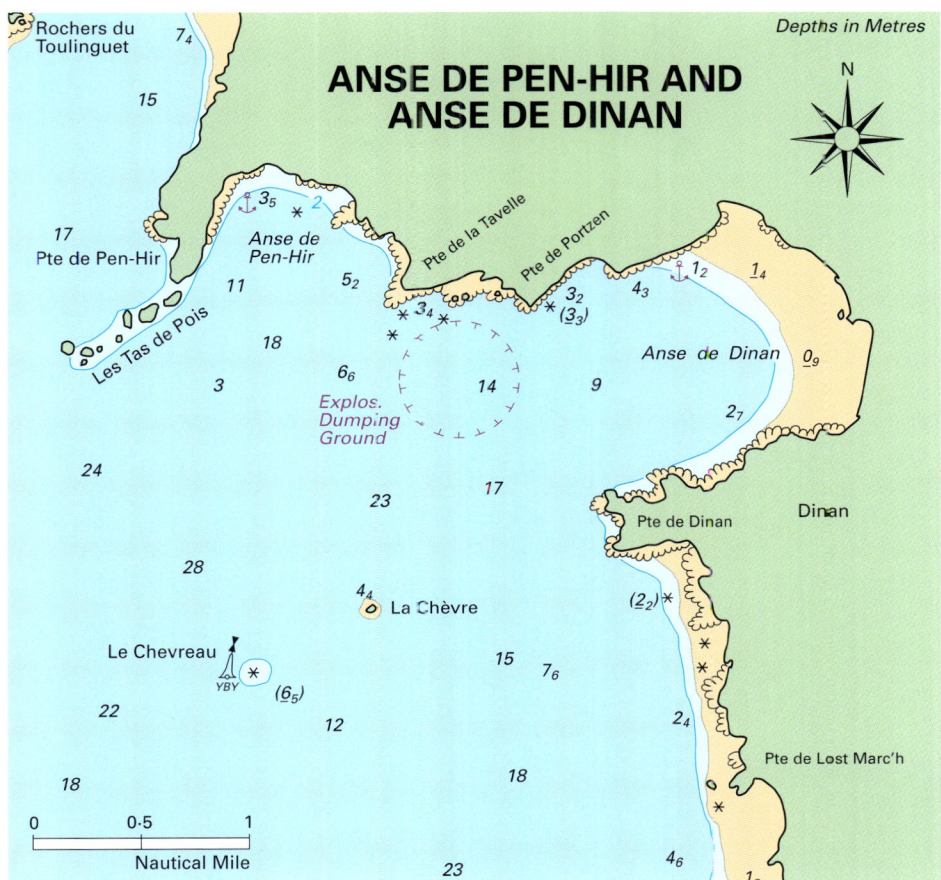

The Anse de Pen-Hir offers good shelter in winds from between west through north to east, although it can be uneasy if there's any onshore swell outside the promontory, because a roll will then find its way in around Les Tas de Pois. Use Admiralty chart No. 2349.

ANSE DE DINAN

A wide shallow bay just three miles east of Les Tas de Pois, the attractive Anse de Dinan offers good shelter in winds from the north and east, but is exposed to any swell from the west. Entry is straightforward from the direction of Les Tas de Pois and the best spot to anchor is usually right in the northeast corner of the bay off the beach.

BAY OF DOUARNENEZ

This magnificent sweep of bay is one of the cruising jewels of Finistère, but is often missed by yachts cruising to a schedule between the Chenal du Four and Raz de Sein. The long, rather austere promontory of La Chèvre, while effectively protecting this splendid gulf from the Atlantic, can also have the effect of discouraging visitors, so wild and harsh can it look in a bit of weather. But entering the Bay of Douarnenez is perfectly simple, and once safely behind Cap de la Chèvre you find yourself in a mini-paradise of a cruising ground, fringed by some of the finest beaches in Brittany. Luxuriant pines flourish around the Crozon peninsula and in a good

SECRET ANCHORAGES OF BRITTANY

MENHIRS OF THE CROZON PENINSULA

Throughout Brittany, but particularly along the coasts, you can often see evidence of the activities of early settlers. The Crozon peninsula is particularly noted for menhirs, those huge standing stones, solitary or in clusters, that look so enigmatic on wild windswept headlands. From out in L'Iroise on a clear day, you can see the rather eerie megaliths high on the cliffs behind Pointe de Toulinguet. It was thought at one time that the mysterious huge stones in lines, circles or standing alone were Druidic, but it's now known that some were erected as long ago as 5,000 BC. When you consider that individual stones can weigh as much as 300 tons and tower up to seven metres high, the scale of engineering involved in their transportation and erection is quite incredible. There are several different formations of stones that must have served different purposes. Menhirs ('*men*' meaning stone and '*hir*' meaning long in Breton), are generally individual upright stones. Dolmens ('*dol*' means flat) often consist of horizontal stones or smaller stones with a large flat stone on the top. These were probably once covered with earth and used as burial chambers. Groups of standing stones are called *alignements* and one theory is that these were used in making astronomical predictions.

Just west of Camaret you can see a sizeable grouping of stones open to the west, the rather mysterious *alignements de Lagatjar*. At first sight they may appear to be a random clustergroup, but it has been shown that these *alignements* were carefully positioned, possibly by a cult that worshipped the sea god or the god of the setting sun. There are also some dramatic megaliths occupying a commanding position overlooking the silver sands fringing the Anse de Pen-Hat. They probably date back to at least 2500 BC (the early Bronze Age), providing an eerie reminder of the vast corridors of time and our own infinitesimal blip on the scale.

Countless competing theories have been advanced for these megaliths, as navigational signposts, primitive astronomical computers, or mysterious phallic offerings. Their great age has certainly endowed them with complex layers of legend but, whatever the original motivation, considerable physical effort and organising ability was needed for their construction.

To reach the Lagatjar site from either of Camaret's marinas, strike inland at the north end of Quai Vauban, fork left along Avenue du Général Leclerc and follow this road southwest until you reach the megaliths on your right-hand side.

CHENAL DU FOUR TO RAZ DE SEIN

summer the area feels more like the old Riviera than Brittany. Soft rolling country encloses most of the bay, rising gradually eastwards to the Parc Régional d'Amorique and the slopes of Ménez-Hom.

The Bay of Douarnenez has two good harbours – Douarnenez itself in the southeast corner and Morgat in the northwest. There are also several interesting anchorages rarely used by visiting yachts, allowing you to get away from the madding crowd even in July and August. Use Admiralty chart No. 2349 for exploring this superb coast.

ANSE DE ST NICOLAS

This somewhat rugged anchorage on the east side of Cap de la Chèvre is a snug spot in northwesterlies. Cap de la Chèvre should be rounded a mile off in a cautious curve and the anchorage approached from the southeast. There are no dangers in the bay and you can tuck close in if the wind is offshore.

The outer harbour at Morgat

SECRET ANCHORAGES OF BRITTANY 153

ANSE DE ST NORGARD

A small rocky bay just over a mile NNW of Anse de St Nicolas, between Pointe de Rostudel and Pointe de St Hernot. There is shelter here in westerlies and northwesterlies, although the coast is rather austere. Approach from the ESE, giving a wide berth to the drying rocks off Pointe de St Hernot.

ILE DE L'ABER

An attractive anchorage two and a half miles east of Morgat entrance, close southeast of Ile de l'Aber and sheltered from between northwest through north to northeast. Coming from Morgat, leave Rocher de l'Aber (21m high) close to port and then turn NNE into the anchorage. Don't fetch up too close to the rocky causeway between Ile de l'Aber and the mainland.

Coming from the south, either from Douarnenez or Cap de la Chèvre, it's usually best to pass east of both La Pierre-Profonde (4.3m high) and Les Verres (9m high) which lie a mile or so SSW of Ile de l'Aber. If passing west of these rocks, be sure to clear Le Taureau (dries 1.8m) which lurks not quite a quarter of a mile north of La Pierre-Profonde.

DOUARNENEZ

There are several anchorages near Douarnenez itself, which can be useful if Tréboul marina or Port du Rosmeur are crowded. In quiet weather or winds from between west through south to southeast, you can anchor in the Rade du Guet, 1–2 cables ESE of Ile Tristan. At neaps, the west side of the Anse du Ris offers reasonable shelter with the wind from between west through south to east. In quiet weather or southerlies, there's a pleasant anchorage off Les Sables-Blancs, a little way west of Pouldavid river mouth. Fetch up about one and a half cables southeast of Rocher Coulinec, or a bit closer to the beach near neaps.

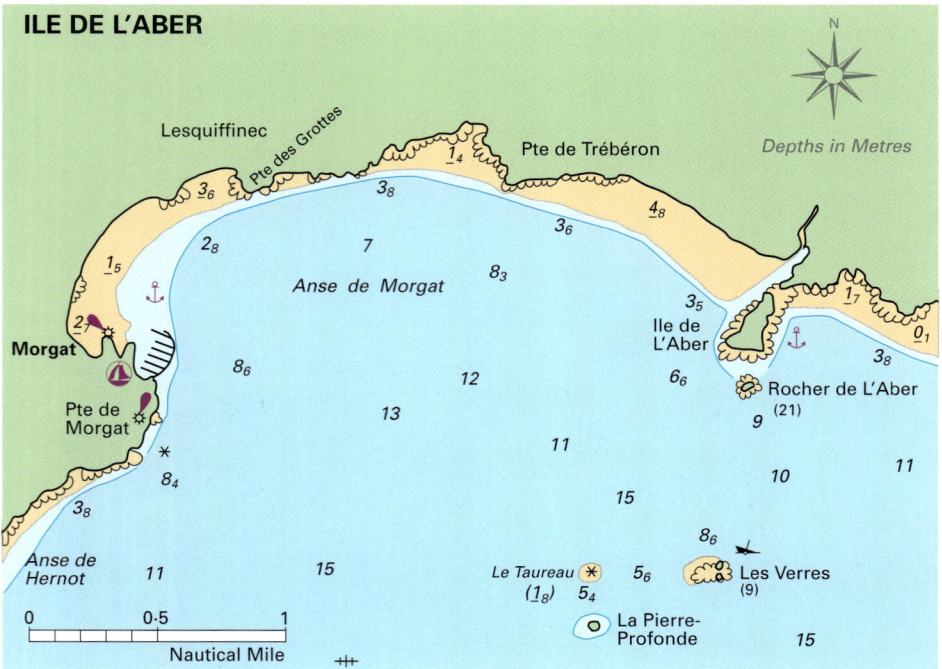

CHENAL DU FOUR TO RAZ DE SEIN

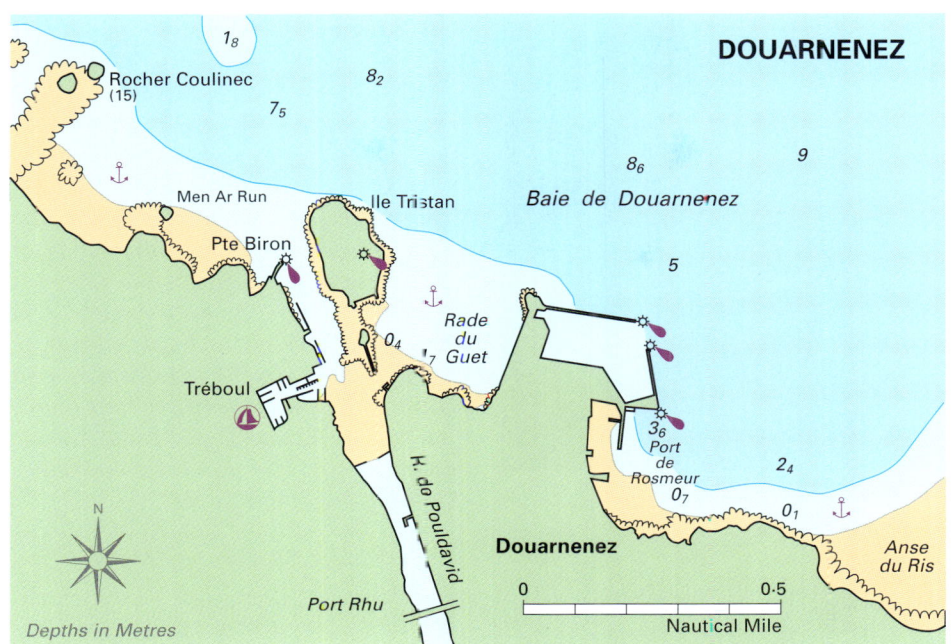

Douarnenez

SECRET ANCHORAGES OF BRITTANY 155

The quiet reaches of the
Pouldavid river

Douarnenez waterfront

Cap de la Chèvre

BAIE DES TRÉPASSÉS

A wide shelving bay just north of the promontory that terminates in the Pointe du Raz. There is a fair weather anchorage on the north side of Baie des Trépassés, about a cable southeast of the islet of Crevendeilet, but keep clear of the fishing boat moorings and keeper-pot buoys. You'll find good shelter here in easterlies, although the bay is wide open to any westerly swell. Trépassés can be a useful spot to wait if you are early on the tide for the Raz de Sein. Approach from due west to avoid the various dangers southwest of Pointe du Van.

USEFUL ADMIRALTY CHARTS

No. 2349 Baie de Douarnenez
No. 3345 Chenal du Four
No. 3427 Approaches to Brest
No. 3428 Brest
No. 3429 Rade de Brest

USEFUL SHOM CHARTS

No. 7121 Baie de Douarnenez
No. 7122 De la Pointe de Saint-Mathieu au phare du four – Chenal du Four
No. 7123 Ile Molène – Ile d'Ouessant – Passage du Fromveur
No. 7400 Rade de Brest
No. 7401 Accès à la Rade de Brest
No. 7423 Raz de Sein

CHENAL DU FOUR TO RAZ DE SEIN

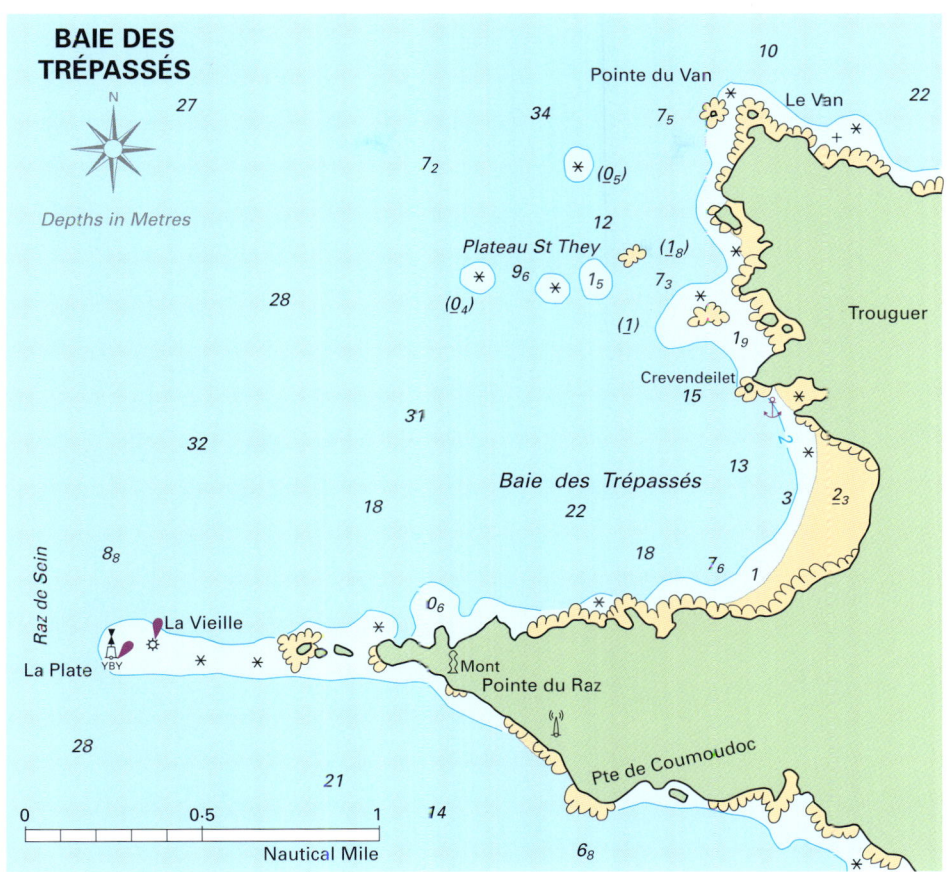

Baie des Trépassés

SECRET ANCHORAGES OF BRITTANY 157

The French sail training ship La Belle Poule in Douarnenez Bay

CHAPTER 5

RAZ DE SEIN TO BAIE DE LA FORÊT

There is a stretch of cruising 'no-man's-land' along the west coast of Brittany, which is definitely well removed from the English Channel and the rocky Côtes du Nord but not yet within the warmer, softer waters of French Biscay. You seem to enter this transitional area at the Raz de Sein, and then hang in limbo for 20–30 miles until Pointe de Penmarc'h is safely astern and you are anchored or moored somewhere in the friendly Anse de Bénodet. In this chapter we negotiate the Raz de Sein and take a quick look at two anchorages off Ile de Sein, before lingering in no-man's-land for a while. Then, having rounded Penmarc'h and entered the Bay of Biscay proper, one can look forward to the various anchorages in the Anse de Bénodet, including the fascinating nooks and crannies around Iles de Glénan.

But first the notorious Raz de Sein, an important milestone in a passage to or from the Bay of Biscay. This uneasy and temperamental stretch of water represents both a natural and a psychological gateway between 'Channel' cruising and the warmer promise of the south. The Raz is also the subject of generations of club-room tales which, apocryphal or not, help to maintain its somewhat sinister reputation.

The Raz de Sein is just one and a half miles wide, from the twin towers of La Plate and La Vieille across to the eastern fringe of the dangers bordering Ile de Sein. Although the tidal stream is fairly weak to the north and south of the Raz, it is powerful in the narrows. The

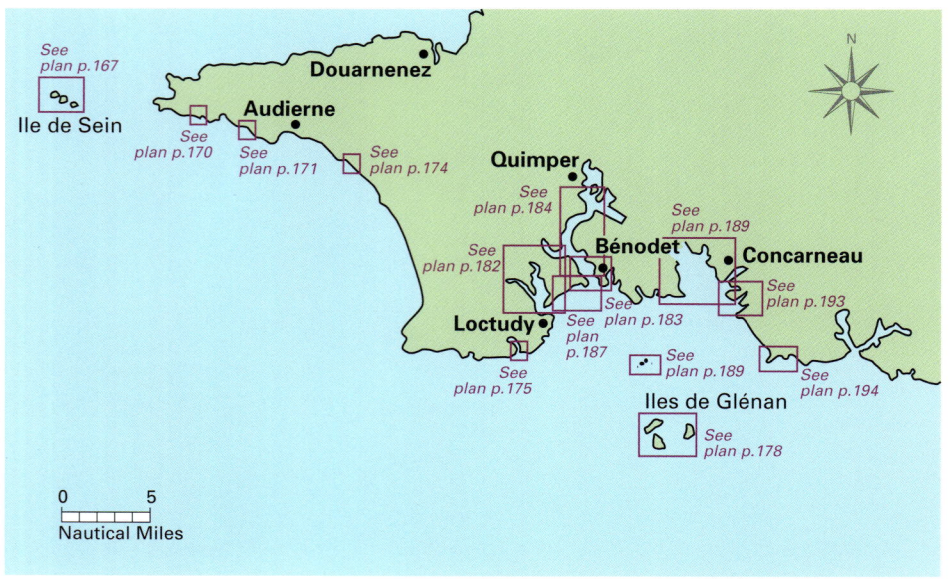

Raz de Sein with Ile de Sein in the distance

strength and concentration of this stream, as it pours over an uneven seabed, can result in a savage maelstrom of overfalls, especially if the tide is weather-going.

As a general rule, therefore, you should take the Raz at slack water, preferably as the stream is about to turn in your favour. Coming from the north, say from Camaret, Morgat or perhaps direct from the Chenal du Four, you would aim to be one and a half miles due east of Tévennec islet half an hour before HW Brest. Coming from the south, from Audierne or Pointe de Penmarc'h, you should be one and a half miles south by east of La Vieille light-tower at five and a half hours after HW Brest.

Slack water is equally important for the approach to Ile de Sein, so that you are not having to cope with fierce cross-tides while trying to pick up the leading marks. I prefer high water slack, because many of the isolated heads are then safely covered. There is, however, a counter argument for approaching at low water slack, when you can see many of the drying rocks and sound into the anchorages knowing the minimum depth of water.

Even today, the remote character of Sein feels worlds apart from the Brittany mainland that is so physically close. Home to about 300 people, the island covers barely half a square mile and is nowhere more than 20 feet above sea level. Most of the houses are on the east side near the harbour, huddled only a yard or two apart against the unrelenting wind. The ageing population is supported by the French department of pensions, by some fishing for the pot and by the tourists who come across by ferry from Audierne to sample a few hours of isolation before returning to mainland comforts.

Few crews on passage through the Raz de Sein are aware of this tiny offshore community, yet the island is easy enough to visit on a quietish day. You just have to know it's there and make a positive decision to divert westward a couple of miles to find one of the most fascinating harbours on the west coast of France. Approaching the Raz from Brest or Camaret, you would normally pass about one and a half miles east of Tévennec rock and its quaint lighthouse before heading SSW to round the twin towers of La Vieille and La Plate. To reach Ile de Sein instead, curve south-about Tévennec, keeping at least a mile off the rock as you round it. Then, if it's high water slack, head for the outer red beacon tower of the island's Chenal Oriental and follow the marks west towards Sein harbour. At low water it's better to make for the green buoy leading to the northern Chenal d'Ezaudi, which has plenty of depth at all tides.

The harbour is on the east side of the island, partly enclosed by a line of rocks and two surprisingly long breakwaters. A narrow peninsula curves out on its south side. The inner part of the harbour dries at springs, although at dead neaps boats of modest draught can find a spot to lie afloat. The outer harbour has water at all tides and several sturdy visitors' moorings, but is more exposed, especially at high water. Light settled westerlies are good for visiting Sein, when the island itself provides reliable shelter.

When the anchor is holding and you gaze around, the overwhelming sensation is one of disbelief that a real community exists out here, unknown and invisible to the stream of boats trekking north and south each season through the Raz de Sein. The village looks neat and well kept from the anchorage, and in warm summer weather the waterfront is as picturesque as many small Brittany harbours. The cottages have traditional high-pitched roofs and painted shutters, and the quayside cafés sport coloured umbrellas. But when the sun goes in, you notice a weathered drabness caused by long exposure to the sea. The buildings stand cheek by jowl, partly because there's very little space but also to keep out Atlantic winds that have a free rein on this low-lying island.

You can land at a slip near the whitewashed tower of Men Brial, the harbour light that shines safe sectors through the rocky dangers fringing Sein. Just north of Men Brial is the pier where the ferry from Audierne usually comes alongside. Looking back towards the mainland from Sein harbour, you'll be able to see boats passing through the Raz de Sein, but none of the crews will see *you* watching them. Ile de Sein remains a very secret place.

The west part of the island is rather desolate, a wild expanse of dunes overshadowed by the lighthouse, some power station buildings and a lonely chapel. You can watch the swell breaking ominously over the rest of the Chaussée de Sein, traditionally known to English mariners as 'The Saints'. This long tail of reefs stretches west for another eight miles as far as Roche Occidentale. Five miles out is the Ar Men tower, whose construction on the outer above-water rock took the local fishermen 14 years of toil before it was completed in 1881.

Back on the mainland, the coast trends more or less due east for eight miles or so from Pointe du Raz towards Audierne. I've included two anchorages along this stretch and a third a few miles beyond Audierne at Pors-Poulhan. There is no practical scope for anchoring off the long southward sweep down to Pointe de Penmarc'h and I have eschewed any rock-dodging in the immediate vicinity of this rather bleak headland. The next feasible anchorage is

Bénodet River and the main lighthouse

in a small bay off a sandy beach just east of Lesconil harbour, a pleasant spot about five miles before the entrance to Loctudy. Loctudy itself, three miles southwest of Bénodet entrance, is accessible at almost any state of tide. This unspoilt river is well filled with moorings and now has a marina, but there is still room to anchor in the shallow upper reaches.

For many yachtsmen, the charming cruising area contained by the triangle of Loctudy, Concarneau and Iles de Glénan, seems to capture the very essence of the South Brittany coast. As you round austere Penmarc'h and follow the offshore buoys east towards the Anse de Bénodet, the whole mood of the seascape softens, the sun usually comes out, and you begin to feel properly on holiday. Beacon towers pop up all over the place, inviting you to choose from several attractive ports of call. The fine sandy beaches are a welcome sight after Penmarc'h and the cluster of low islands to starboard, the inimitable Iles de Glénan, holds a promise of fascinating anchorages yet to be explored.

Iles de Glénan are home to the legendary sailing school, the Centre Nautique des Glénans, which was established on Ile Cigogne shortly after the last war. This was a far-sighted project with the aim of encouraging self-reliance through seamanship in the country's up and coming generation. Glénan boats roam far and wide, and I well remember the tiny but functional aluminium sloops, without engines or electronics and packed with six or more crew, that used to tack into Dartmouth after a 48-hour passage direct from Glénan.

Half a dozen miles east of Anse de Bénodet is a pronounced inlet known as Baie de la Forêt, with the colourful port

BUILDING THE AR MEN LIGHT

The long tail of reefs known as Chaussée de Sein straggles out into the Atlantic for eight nautical miles beyond the west end of Ile de Sein. In the days of sail and steam, many ships came to grief on these treacherous shoals, which remained unguarded and unlit until 1881. The project to build a lighthouse on the Chaussée de Sein was mooted as early as 1829, mainly because of the growing importance of the new passenger and packet service to America. Because the French ships on this route called at both Le Havre and Brest, they often had to pass Chaussée de Sein fairly close. As planning for a new lighthouse continued, there was much debate and uncertainty about where best to try and build a substantial stone tower. The only rocks that were above water at all tides were too far into the long chain of reefs to be useful. A new lighthouse only halfway along could actually be misleading and potentially more dangerous than not having a light at all.

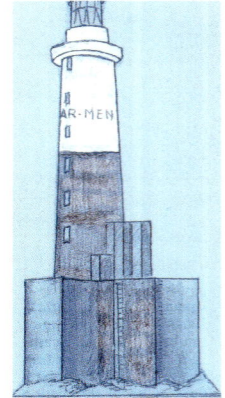

This complex problem was given careful thought by the engineers of Les Phares et Balises, the French government department responsible for lighthouses and navigation marks. They realised that, to be worth the cost of construction, any new lighthouse on Chaussée de Sein should stand as far seaward as possible, on one of the outlying rocks. After years of discussions and surveys, attention turned to what seemed like the only possible contender – the Ar Men rock, which uncovered for barely two hours a day during periods of spring tides. Much thorough research was conducted and analysed before it was finally decided that the Ar Men was a viable site. However, even once this decision had been reached, it took another 14 years to build the lighthouse because of the sheer difficulty and danger involved in working on the tiny, precarious platform of rock. Not only did the tides have to be just right, unsuitable weather often prevented access for days, sometimes weeks at a time. The fishermen of Ile de Sein eventually agreed to act as labourers and boatmen, although initially they weren't at all keen to be involved. They were finally persuaded by a deal in which Les Phares et Balises agreed to rebuild and improve the breakwaters around Ile de Sein harbour. Work started on Ar Men in 1867. With only one and a half metres of rock above water level at the lowest tides and only a few square metres of area to work on, it took many visits just to make the first 15 holes for the iron reinforcing bars that would tie the foundations together. In those days of sail and oar, just reaching and landing on the rocky plateau was a feat of courage and seamanship in itself. The workers wore large cumbersome life jackets and were attached by rope to prevent them being washed off the rock.

During the whole of the second year, another 34 foundation holes were chipped out in just 18 hours of effective work. The first real construction began in 1869, when the masons managed to put in 44 hours of work and build 25 cubic metres of masonry covering the rock. English Portland cement was found to be best for the job, curing quickly and strongly in the difficult conditions. By the end of 1877 the tower had risen 16.5 metres above the rock, although this was barely 12 metres above high water level. However, work continued slowly but surely over the next four years and the light was first illuminated at midnight on 30th August 1881.

Even after this long but successful construction, the Ar Men lighthouse remained one of the most exposed and difficult to service of all the lights around the Brittany coast. The keepers could be cut off for weeks on end by weather or swell, having to rely on the provisions and water in the lighthouse store. The Ar Men light was automated on 10 April 1990.

Raz de Sein and Pointe du Van

of Concarneau in its southeast corner. At the head of the bay, a narrow dredged channel leads up a shallow creek behind Cap Coz to Port la Forêt yacht harbour. You can anchor off the entrance to this channel, just south of Cap Cos, and at Beg Meil off the southwest side of Baie de la Forêt. Opposite Beg Meil, on the Concarneau side of the bay, there are quiet spots to anchor in the Anse de Kersos, a peaceful and attractive inlet half a mile south of Concarneau harbour entrance. There are anchorages, too, as you start cruising southeast out of Baie de la Forêt and then eastwards towards the picturesque Aven and Bélon Rivers.

ILE DE SEIN

Ile de Sein is not difficult to approach at high slack water, provided you set careful waypoints to lead safely to the outer end of your chosen approach channel. The main ways in are Chenal d'Ezaudi from the north and Chenal Oriental from the east. The trickier Chenal d'Ar-Vas-Du is best left to the locals.

Chenal d'Ezaudi This 'all tides approach' has a corridor of deep water even at low springs. Chenal d'Ezaudi is usually the best way in if you arrive off the Raz at slack water low. A green buoy marks its outer end, left about 75m to starboard as you approach from the north. You then leave Ezaudi above-water rock a little over 150m to starboard. A useful steering line keeps Guernic green beacon just open to the east of Men Brial lighthouse at 187°.

Peacefully at anchor off Ile de Sein

Chenal Oriental This route is used every day by the ferry from Audierne. From the Raz de Sein, with La Plate and La Vieille bearing just south of east, simply head west to keep Cornoc-Ar-Vaz-Nevez red beacon tower reasonably fine on the port bow. As you draw nearer, steer to leave this tower 120m to port and then Penbara above-water rock a short cable to starboard. By then you should see the clear leading marks for the Chenal Oriental:

Front mark – a white tower on the north end of the line of rocks protecting Sein harbour from the east.

Rear mark – a large white pyramid (Plas ar Scoul) on the west end of the island, with a prominent orange top.

Follow this transit at 265° to within 200m of the front white tower before curving north-about the rocky point to enter the harbour.

ILE DE SEIN ANCHORAGES

The most sheltered anchorage is towards the south part of the harbour, which dries at datum but retains enough depth for moderate draught boats to stay afloat or just nudge the bottom near neaps. At springs you have to fetch up more or less opposite the main ferry jetty and Men Brial harbour lighthouse in order to stay afloat. This latter position is rather exposed to the east above half-tide, but is well protected from the south and west, and in any case increasingly snug as the ebb falls away.

In quiet weather or in any winds from the south, you can anchor off the north side of the island near the main lighthouse, a remote spot that captures the wild atmosphere of the island perfectly. This anchorage is quite easy to reach either from the harbour, or as you come in via the Chenal d'Ezaudi or Chenal Oriental. From a position about 300 metres NNW of Nerroth islet and its white beacon, bring Ile de Sein main lighthouse bearing due west true and approach it on this line until Conolloc-Est red beacon tower is bearing due north true. Then edge southwest towards the beach and fetch up in just over a metre LAT.

You should move round into the harbour if the wind swings out of the south or a westerly swell sets in, but in quiet weather I have swung here overnight, under the four eerie sweeps of the lighthouse. You can leave this spot fairly easily after dark by reversing the approach directions and keeping the lighthouse bearing due west true astern until you pick up the northern white sector of Men Brial lighthouse, which you can then follow just west of south into the harbour.

RAZ DE SEIN TO BAIE DE LA FORÊT

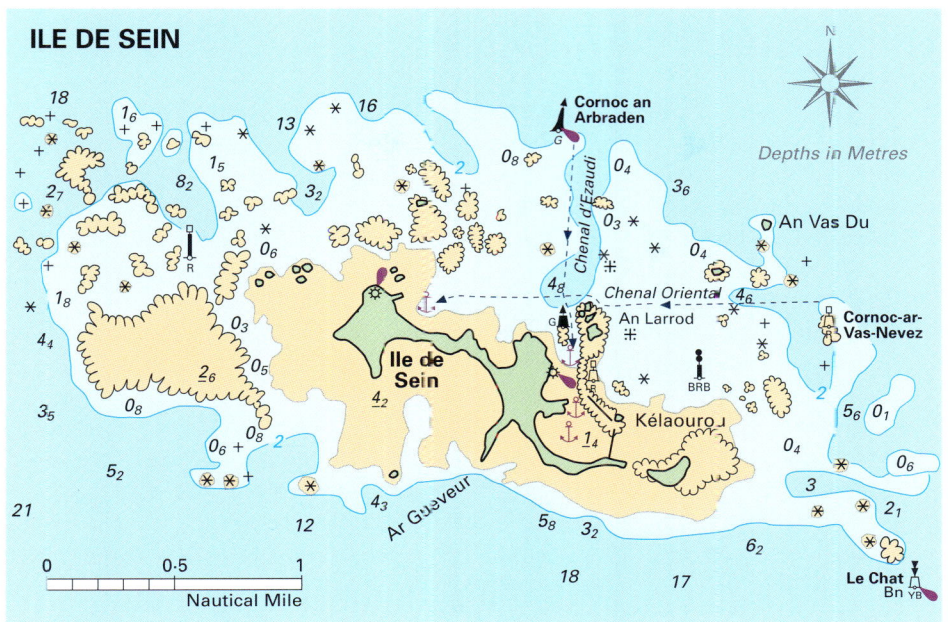

Ile de Sein

SECRET ANCHORAGES OF BRITTANY ⚓ 167

THE EERIE RAZ DE SEIN

The Raz de Sein has a legendary reputation and even the hardiest French yachtsmen usually mention this powerful tidal gate in hushed tones. This is understandable to anyone who has caught this stretch of water in an irascible mood, when in extreme conditions the Raz is a devil's cauldron of turbulence, overfalls and chaotic breaking water. At the peak of the tide, La Vieille lighthouse and La Plate beacon tower seem to be under way, like the bizarre conning towers of submarines gamely stemming a six-knot stream, but going nowhere.

Yet on a quiet summer day at slack water, the Raz can look as innocent as the Serpentine. More often than not, you find yourself motoring through with the mainsail limp and the sea glassy calm, with only a few tell-tale swirls to hint at the potential for mischief when the stream gets moving again. Then, as you wonder what all the fuss was about, you can pass La Plate a bare cable off, with crowds of French sightseers clearly visible up on Pointe du Raz. A few miles to seaward, the low profile of Ile de Sein barely lifts above the sea, unreal with its tall lighthouse and clusters of remote cottages. It's difficult to imagine that about two hundred people live on this tiny island all year round. The Raz de Sein is barely two miles wide and represents the last real tidal hurdle for yachts cruising down to south Brittany and beyond. On the west side are various beacon towers guarding the plateau of dangers around Ile de Sein. On the east side, the elegant stone lighthouse of La Vieille and the W Cardinal tower of La Plate mark the end of a line of rocks straggling well out from Pointe du Raz. A curious feature of this temperamental passage is that just to the south, in the Bay of Audierne, and just to the north in L'Iroise, the tidal streams are generally rather weak. Things only become frantic in the Raz itself, where streams reach six knots at springs.

You can appreciate this phenomenon from a small-scale Biscay chart, which shows the Raz as a tiny gap at the top of the Bay. Most of the French Biscay coast has quite weak tidal streams, which tend to set onshore during the flood and offshore during the ebb. Because this coast faces the main Atlantic tidal wave squarely, there's no great scope for tidal acceleration except at certain strategic narrows. On this grand Biscay scale, the Raz de Sein is the obvious point for leakage. The immense oceanic tidal wave surges into the Bay twice every 24 hours, filling up the rivers and inlets along the coast but able to escape at the north edge through the narrow strait between Pointe du Raz and Ile de Sein. As the giant wave recedes and the ebb begins, the tidal vacuum sucks water back down through the Raz from the English Channel, via the Four and L'Iroise. But because there are wide bays on each side of the Raz, the fast-flowing stream is very localised, as with a tap emptying into a large, deep bath. You don't have to drop more than a couple of miles south of La Vieille and La Plate to reach the more sluggish tides of the Bay of Audierne. Heading north, the effect is similar. Once clear of tiny Tévennec island and Pointe du Van, you have escaped the clutches of the Raz and reached the languid waters of L'Iroise.

RAZ DE SEIN TO BAIE DE LA FORÊT

Ile de Sein harbour

Ile de Sein lighthouse

SECRET ANCHORAGES OF BRITTANY ⚓ 169

Approaching Ile de Sein

Below: Waterside houses on Ile de Sein

Bottom of page: Anse de Feunteunod

ANSE DE FEUNTEUNOD

A small cove on the south side of the Pointe du Raz, not quite three miles along the coast from La Plate beacon tower and just east of Pointe de Feunteunod. Approach from due south at any state of tide, having rounded Pointe de Feunteunod at least two cables off if coming from the Raz. There is good shelter from WNW through north to east if you edge well into the cove. If the wind shifts while you are there, Audierne is only seven miles away to the east.

ANSE DU CABESTAN

Anse du Loc'h

This wide sandy bay lies four miles east along the coast from the Anse de Feunteunod and only three miles from Audierne. It makes a pleasant anchorage in settled weather when the wind is from anywhere between north and east. Approach from the southwest at any state of tide, making sure that you avoid Basse du Loch (dries 2.1m) and Rocher Sud de Porz-Tarz (dries 3m) if you are coming along the coast from the direction of the Raz. St Tugen church spire is a useful landmark behind the beach.

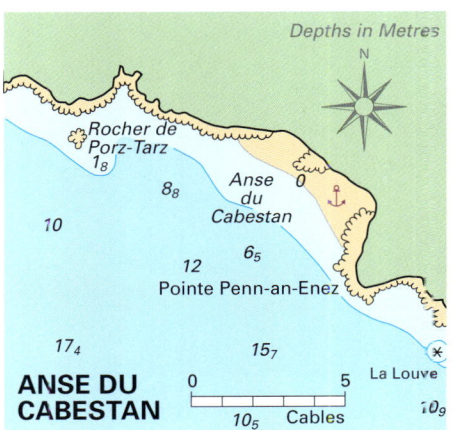

MOULES AU CIDRE

Mussels are quick and easy to cook on board if you get a chance to buy some in a local market, and for *moules au cidre* you need a bottle of medium or slightly sweet (*douce*) Brittany cider. For a good syrupy sauce it's best to start reducing the cider slowly before doing anything else, so pour rather more than half your bottle into a wide pan, bring the cider to the boil and then bubble briskly while you are preparing the mussels.

First make sure the mussels are clean on the outside, by scrubbing the shells with a kitchen brush and rinsing them in clean water (not marina water). Also gently pull off any of the 'beard' which often attaches between the two halves of the shells. Throw away any mussels that stay open as you scrub them, as they'll probably be dead and therefore risky. (It only takes one dodgy mussel to bring the day to a gloomy conclusion.)

In a decent-sized pan or casserole, melt a generous chunk of butter and gently fry three finely chopped shallots (or a medium-sized onion) and a couple of chopped cloves of garlic. Stir in a little freshly chopped parsley and a pinch of thyme before adding the mussels, then pour in as much of your remaining cider as it takes almost to cover them. Grind over some black pepper and cook at a fairly high heat until the liquid just starts to bubble and the mussels open – but no longer. Remove the mussels from the pan (discarding any rogues that are still closed) and put them in a bowl that has been warmed. Then strain the cooking liquor into your already bubbling cider, which ideally by now has been reduced to a *quarter of its volume*. Stir up this combined sauce well, bubble vigorously a little longer to reduce even further and then pour what should be a fairly syrupy liquor over the mussels. Add a new generous sprinkling of

chopped parsley, stir the mussels in the bowl to circulate the liquor and parsley and then serve. A bottle of well-chilled dry Brittany cider is the perfect accompaniment.

THE TRAGIC TALE OF YS

Throughout western Europe and in the long folk history of coastal countries everywhere, there are stories of lost towns and cities under the sea. Sailors and fishermen tend to be superstitious people and many have told of hearing church bells under the waves when the wind and tide are in certain combinations. Somewhere off the timeless west coast of Brittany, calm and still under the restless waves, stands the ancient city of Ys. Some say that Ys might be found off the rugged Baie des Trépassés, just north of the mysterious Raz de Sein. Others argue that Ys has a less tempestuous location in the Bay of Douarnenez, and at exceptionally low tides the shadowy outlines of walls and stones can be seen.

The story goes that Princess Dahut was a cruel and wanton daughter to King Gradlon, a noble and blameless ruler. The capital of Gradlon's kingdom was the city of Ys, which was built on the west Brittany shore below sea level. To protect the city there was an immensely strong wall with sluice gates and only the king had the key to these gates. The inhabitants of Ys grew dissolute and immoral, following the example set by Princess Dahut rather than that of her good father. Despite frequent appeals from her father and stern admonishments by St Gwennolé, Dahut would not mend her ways.

God determined to punish the city and its inhabitants. The Devil came to Ys in the guise of a handsome young man, who seduced Dahut and persuaded her to give him the key to the gates. She stole the key from her father and then the Devil opened the sluices and let the sea pour into the city. St Gwennolé warned King Gradlon just in time for him to escape. He mounted his horse and, gathering Dahut up with him, tried to ride away from the incoming water. But because Dahut was with him, the king couldn't ride fast enough to escape the rushing sea and St Gwennolé ordered Gradlon to cast his daughter off into the water. When he finally did this, his gallant horse was able to outrun the rising sea level and Gradlon managed to reach dry land safely.

The city of Ys still lurks below the waves off the coast of Brittany and many sailors and fishermen have heard the cries of its inhabitants and the mournful tolling of bells. You may hear them too, particularly on a grey day when westerly weather is sending long Atlantic seas onto that wild unchanging coast between Pointe de St Mathieu and the Raz de Sein.

King Gradlon

PORS-POULHAN

Pors-Poulhan light

A tiny pier harbour for small local boats, three and a half miles ESE along the coast from the entrance to Audierne. You can anchor off the entrance to Pors-Poulhan in northeasterlies and it's not far to Audierne or Sainte Evette if the wind should shift.

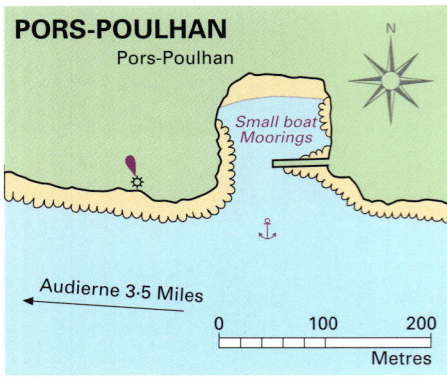

ANSE DE LESCONIL

Lesconil is a small but active South Brittany fishing harbour seven miles east of Pointe de Penmarc'h. From June–Sept visitors can lie afloat in the inner harbour on buoys or at pontoons in the northeast corner, but there's an outer anchorage 3–4 cables ENE of the pierheads, sheltered in any winds from between west through north to northeast. The easiest approach is from

Pors-Poulhan entrance

RAZ DE SEIN TO BAIE DE LA FORÊT

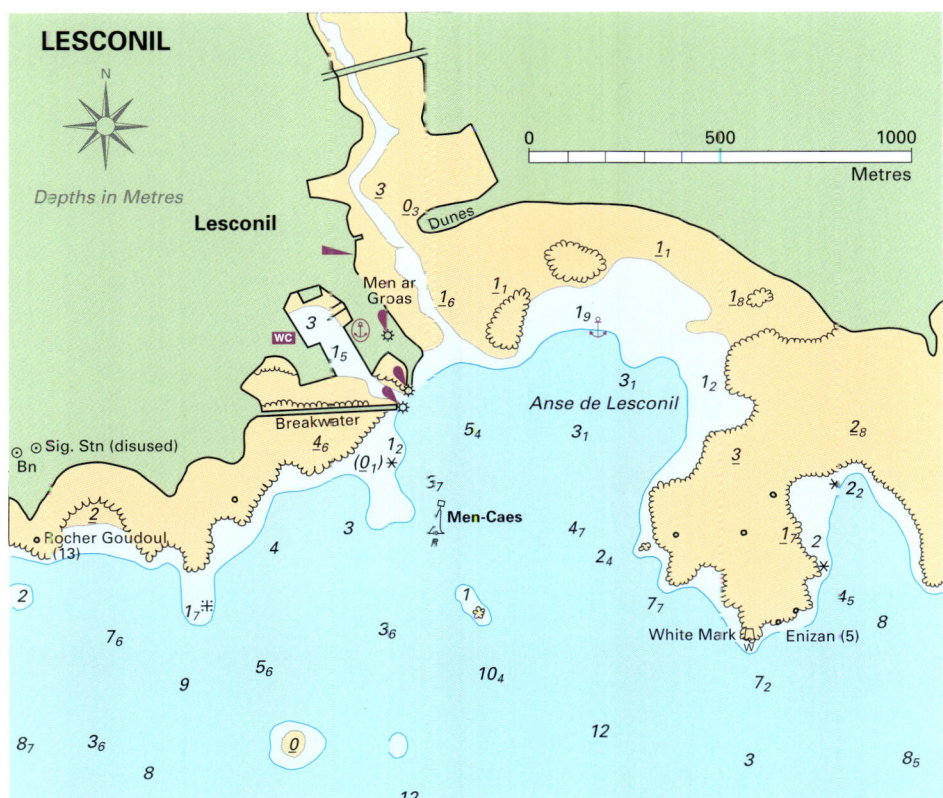

Pointe de Penmarc'h

SECRET ANCHORAGES OF BRITTANY

FISH HUNTERS OF LE GUILVINEC

Cruising down towards the Biscay coast of France, perhaps on passage from the irascible Raz de Sein towards the low promontory of Penmarc'h, you'll often come across large powerful fishing boats heading straight offshore with an almost tangible sense of commercial purpose. Fishing is a serious business around the whole of France, but the modern fish-hunting ships that prowl the deep waters of Biscay are painstakingly designed to maximise catches and equipped with the very latest technology to that end. In the days of sail, it may have been romantic to see *bisquines* and luggers working gently offshore in a Biscay sunset, but there is something altogether more sinister about watching the powerful fleets setting off now to the combined growl of thousands of horsepower. Around Penmarc'h there are several large bases for these high-tech fish-hunters – Le Guilvinec, St Guénolé, Lesconil, Loctudy – but Guilvinec, about four miles east of Eckmühl lighthouse, always seems to me to have the most intense motivation for the gruelling business of catching fish. The approach to Le Guilvinec is not too difficult in fair weather, but looks austere and somehow tricky from the offing, as if this location for a modern fishing harbour had been chosen to discourage all but those professional seafarers engaged in the industry. Visiting yachts are accommodated at the head of the long snug harbour, but you feel rather dilettante amongst all those hardy fishing crews and their massively expensive working ships.

Yet it's interesting that the skippers of the largest high-tech vessels operating out of Le Guilvinec rarely arrive on the quayside dressed in oilskins, sou'westers and sea boots. They come aboard more casually and dapperly dressed in an almost corporate style, probably with a leather briefcase. Their wheelhouses have the feel of modern offices – spacious, air conditioned and bristling with computer screens and instruments. But anyone cruising in France soon appreciates why fishing is so highly organised. Every day of the year, in every town and village within at least 50 miles of the coast, you can see fresh fish of all kinds on every restaurant and bistro menu. There it all is, waiting to be cooked and eaten, and someone has had to range far and wide in all weathers aboard sophisticated vessels to catch it. Fishing is a hard, dangerous and chancy occupation and, in a way, it always amazes me that fish is not more expensive than it is.

Pointe de Penmarc'h

Lesconil

Karek Greis E Cardinal buoy, thence making good 335° for just over one and a half miles to leave Enizan rock (5.5m high) a quarter of a mile to starboard and Men-Caës red beacon tower one and a half cables to port. Once past Men Caës, turn northeast to enter the anchorage, but be sure to avoid the extensive ledge of drying rocks on the east side of the bay.

ILES DE GLÉNAN

This fascinating archipelago is a must on a South Brittany cruise, lying about 10 miles SSE of Bénodet and a similar distance SSW of Concarneau. The world-famous sailing school, the Centre Nautique des Glénans, has bases on four of the main islands – Ile de Penfret, Ile de Bananec, Ile Cigogne and Ile de Drénec. Glénan is a honey-pot for local French boats during summer weekends, so try to call here midweek if possible. Although bordered by numerous rocks and shoals, the islands are not difficult to navigate in quiet weather with good visibility. The best line of approach for strangers is from the NNE near high water, making good about 205° to leave the north end of Ile de Penfret fairly close to port. This is a particularly convenient entrance if coming from Concarneau or La Forêt.

Penfret is the easternmost island of the group, easily identified by the prominent lighthouse (white with a red top) on its north side. It is also the main base. Coming in past the west side of Penfret, you should steer towards the black and white beacon on Ile Guéotec. Once halfway along Penfret, you are inside the archipelago and can follow the chart to whichever anchorage seems best for the conditions.

La Chambre From the west coast of Penfret, steer a shade north of west to keep the houses on Ile St Nicolas fine on the starboard bow, giving a good berth to the rocky shoals that extend southeast from Ile de Bananec. La Chambre is the popular pool just south of Ile St Nicolas and offers fair shelter from between

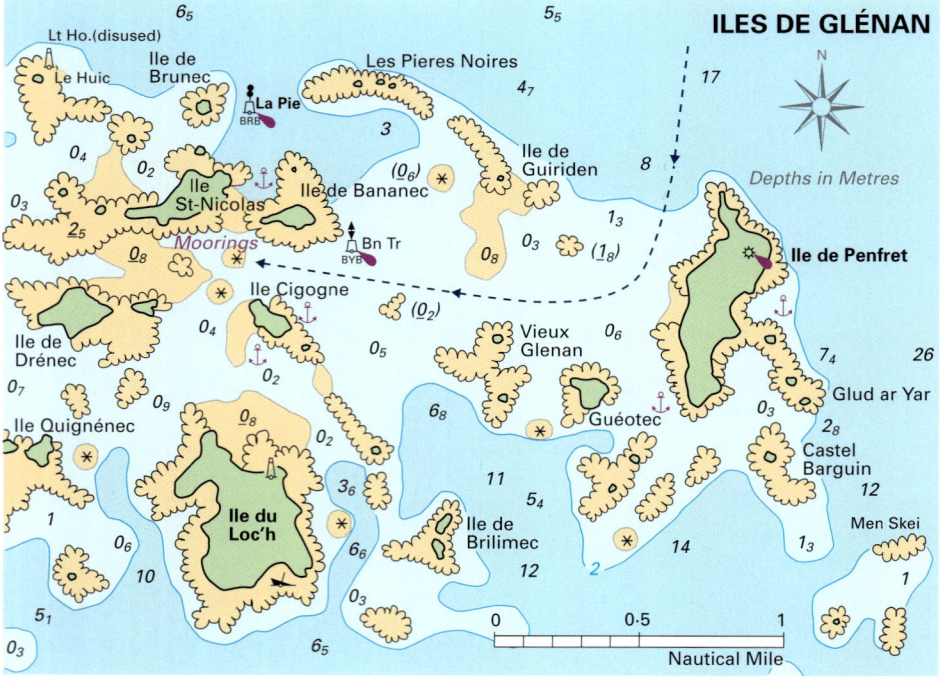

Iles de Glénan

north through west to south. Anchor clear of all the visitor moorings, but don't stay overnight unless the weather is reasonably set fair.

Ile Cigogne Between springs and neaps you can anchor off the east and south-west sides in quiet weather.

Ile de Bananec In moderate winds with any south in them, you'll find a snug anchorage in the moorings bay formed by the east side of Ile de St Nicolas and the ledge of drying rocks that extends north from Ile de Bananec.

Ile du Loch In quiet weather near neaps, you can anchor between Ile Cigogne and the north side of Ile du Loch, opposite and about two cables north of the conspicuous chimney on the northeast corner of Loch.

Ile de Penfret There's a good but often crowded anchorage off the southwest side of Ile de Penfret, outside the local moorings. Penfret protects you from the east and the other islands provide shelter from the west. On the east side of Ile de Penfret is an easy to enter but also often crowded anchorage in the bay just south of the lighthouse, with good shelter from

CENTRE NAUTIQUE DES GLÉNANS

Lying nine miles offshore from Bénodet or Concarneau, the fascinating Iles de Glénan are home to the now legendary sailing school – Le Centre Nautique des Glénans. In 1945 Hélène and Phillipe Viannay visited the islands and were so taken with the archipelago that they returned the following year in their boat. A year later they founded the Centre International de Formation (CIF) and made Ile du Loc'h their first base. The original objective was to help to reintegrate into civilian life young French people who had been involved with the Resistance movement.

By 1952 the CIF had led to the establishment of the Centre Nautique des Glénans and gradually bases were established on all the islands of the group. Students now came from all walks of French life learning to develop self-reliance and social skills through the communal activity of sailing. The central idea of cruising from these Brittany islands aboard small yachts has proved highly successful and has caught the imagination of successive French generations. In the 70 years since it was conceived over 200,000 students of all ages have passed through the Centre Nautique in the islands and three times this number have completed the courses at other locations, absorbing what its founders called that *esprit du projet Glénans*.

Glénans boats continue to roam far and wide and many of the smaller cruising boats have no engines. I well remember the tiny but functional aluminium sloops, without engines or electronics and packed with six or more crew, that used to tack into Dartmouth in the 1960s after a 48-hour direct passage from the islands.

Although bordered by rocks and shoals, the Glénan islands are not difficult to navigate in quiet summer weather with reasonable visibility. You can safely approach the archipelago using the medium-scale Admiralty chart No. 2820, but you should then switch to the large-scale Admiralty chart No. 3640 or the French SHOM chart No. 6648. There are several idyllic daytime anchorages around the islands, although most now become very busy during summer weekends. But if you are cruising this area early or late in the season, you can discover something of the true character of Glénan by anchoring for a couple of nights within the archipelago and absorbing the vibes of the Centre Nautique.

RAZ DE SEIN TO BAIE DE LA FORÊT

Ile de Penfret
Ile St-Nicolas

SECRET ANCHORAGES OF BRITTANY

RIVIÈRE DE PONT-L'ABBE

Loctudy and Ile Tudy are well covered by the pilot books, but you can find a secluded neap anchorage a little way upstream from the main mooring area, off the southeast tip of Ile Chevalier. Fetch up opposite Château de Najac, or edge further north along the island if the depth allows. Good holding in muddy sand.

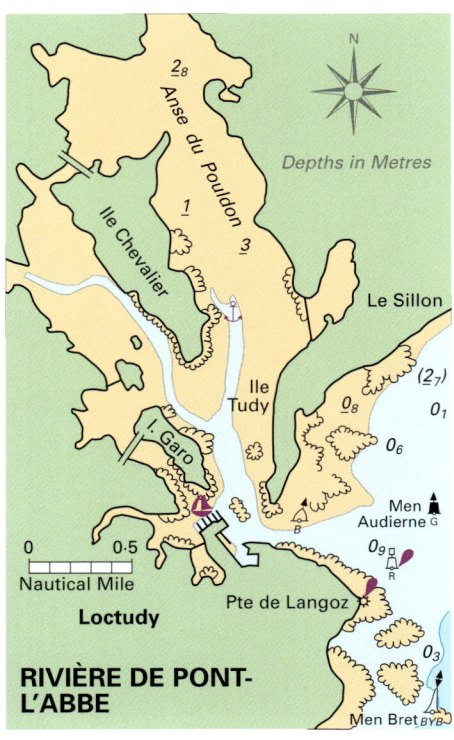

RIVIÈRE DE PONT-L'ABBE

ANSE DU TREZ

This attractive sandy anchorage lies just inside the mouth of the Bénodet River on the east side. Arriving in the Anse de Bénodet near Basse Boulanger S

the west. Approach this anchorage from due east, leaving Kastell Razed islet on the south side of the bay a cable to port.

Approaches to Loctudy

Ile Tudy

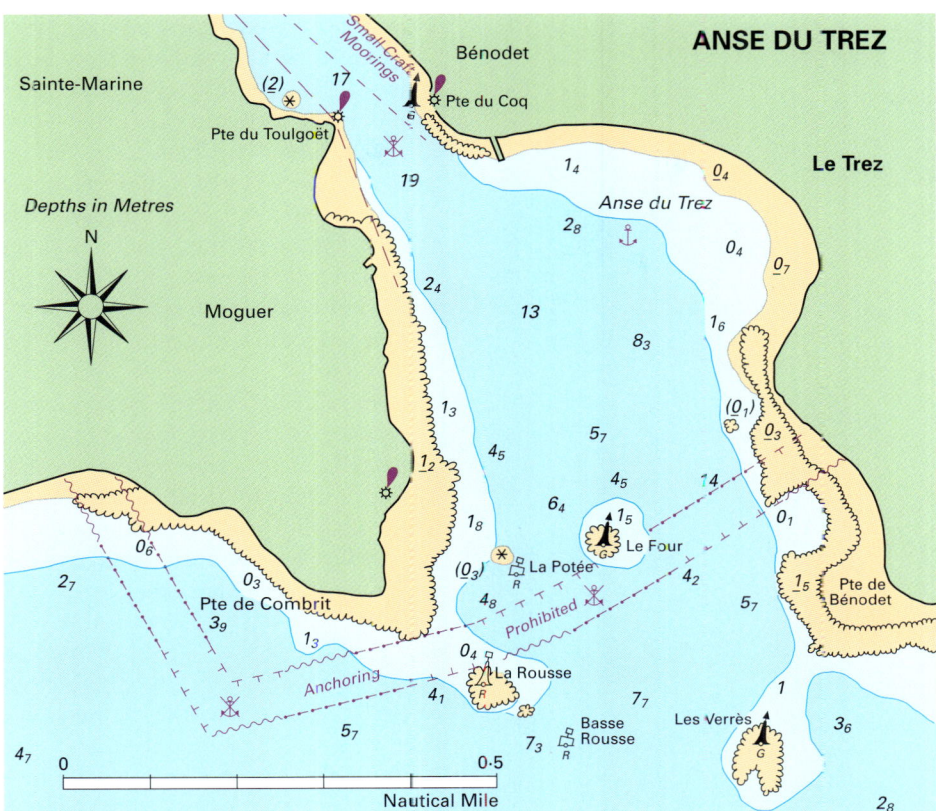

Cardinal buoy, head northeast to pass between the gateway formed by Basse du Chenal E Cardinal buoy and Basse Malvic W Cardinal. From here steer just east of north for Bénodet entrance, passing between Les Verrès green tower and the red buoy close southeast of La Rousse red beacon. Then continue into the river between Le Four green beacon and La Potée red buoy before turning NNE for the Anse du Trez. Anchor about 150 metres from the beach, or closer in if the depth allows near neaps. There are various racing marks in the bay and you'll probably be surrounded by small sailing dinghies during the day, but everything quietens down towards evening.

RIVIÈRE DE L'ODET

Although Bénodet itself is much frequented by visiting yachts, relatively few venture far upriver above the suspension bridge (Pont de Cornouaille). Yet there are several idyllic anchorages in the five-mile stretch between the bridge and the small village of Lanros. A mile or so above Pont de Cornouaille you can anchor off the mouth of the Anse de Combrit. Above Combrit, the main river valley runs north for about two miles before winding off to the east. Just before this change of direction, the Odet cuts between high wooded cliffs known as Les Vire-Court.

Although the east-going stretch is quite narrow, there's plenty of depth if you follow the spars that mark various rocky shoals jutting out from the banks.

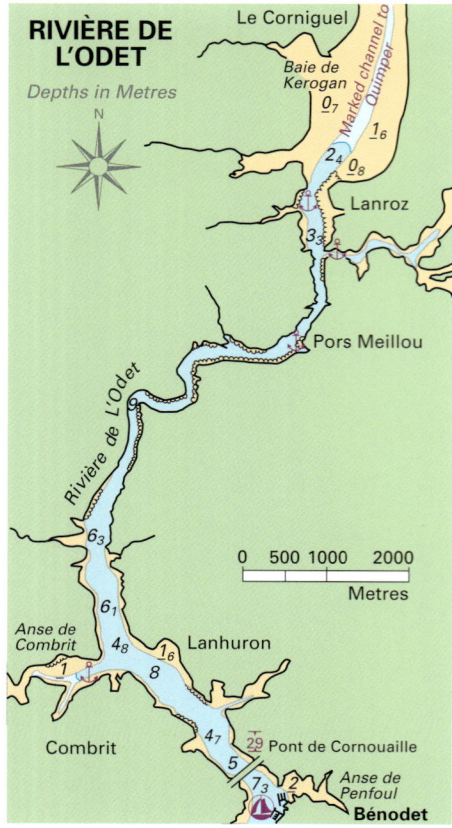

After another mile or so the river turns north again, just beyond a shallow bight known as Pors Meillou.

When anchoring in the upper Odet, edge well into a bay, bight or mouth of an inlet. The river bed is often rocky, so you need to fetch up in those quiet corners, out of the tide, where mud has settled to provide good holding. This also keeps you clear of tourist *vedettes*, and the barges that go up to the wharves at Corniguel.

About five miles above Bénodet, you can anchor off the west bank in the bight opposite Lanroz, again tucking in close. Above Lanroz you need high water for further exploration. The channel across the shallow Baie de Kérogan is marked by red and green spars that lead NNE for just over a mile, before turning west past Port du Corniguel and then north again towards Quimper. The river dries above Corniguel, but it's fun to take the dinghy up to Quimper on the tide.

Anse de Combrit Just over a mile above Pont de Cornouaille on the west side of the river, you can anchor off or just into the mouth of Anse de Combrit, where the holding is good over a muddy bottom. Keep over towards the north

Manoir de Kerouzien on the Odet River

Waterfront café at Sainte-Marine

Odet River

Château de Kerambleiz on the Odet River

side of this shallow inlet when entering or leaving, watching the echo-sounder carefully, especially below half-tide.

Pors Meillou This attractive spot to anchor lies three miles upstream from the Anse de Combrit on the east side of the river. Anchor well into the mouth of the inlet or close inshore just south of it to find the best holding and miss the worst of the tide.

Anse de Toulven Just over half a mile above Pors Meillou, the narrow entrance to the Anse de Toulven opens off the east side of the river, leading to a tranquil anchorage a little way in where the creek widens on its south side. Anchor towards the south shore in between one and two metres, swinging on a fairly short scope. Keep to the north side of the creek as you come in, to avoid a muddy shoal off the south arm of the entrance and a rocky ledge, covered at high water, that juts out from the south bank just before the creek opens out.

RAZ DE SEIN TO BAIE DE LA FORÊT

Lanroz You can anchor in the bight on the west side of the main river, just opposite Lanroz and its château over on the east side. This sheltered spot looks north across a wide expanse where the river opens into the shallow Baie de Kérogan.

PLAGE DU TEVEN

Back out in the Anse de Bénodet, it's pleasant to anchor off the splendid long beach of Teven that stretches southwest from near Bénodet entrance for a couple of miles towards Loctudy. Admiralty

Anse de Toulven anchorage up the Odet River

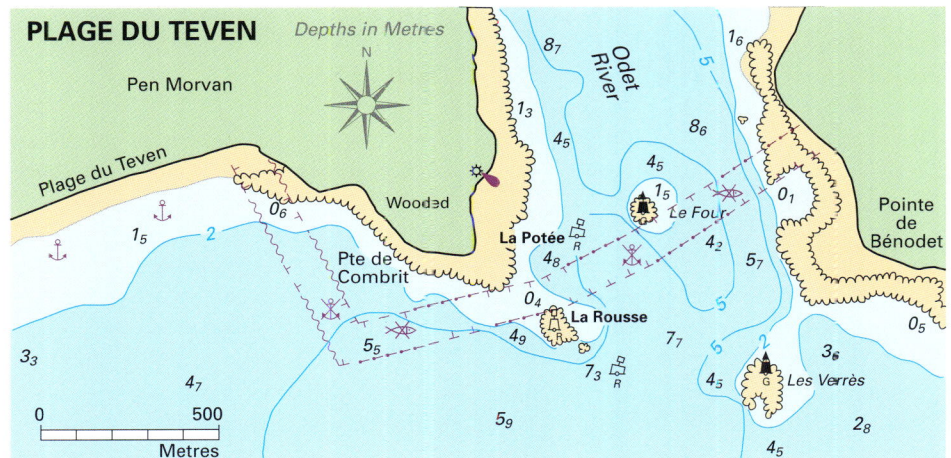

SECRET ANCHORAGES OF BRITTANY 187

Baie de la Forêt

chart No. 3641 shows very thin soundings off this beach, but between springs and neaps, and even on ordinary springs in calm weather, most boats will be able to edge in quite close to the east end of Plage du Teven, within a mile of Pointe de Combrit. In settled northwesterlies you can lie safely here overnight, although a slight roll usually finds its way into the Anse de Bénodet to disturb your sleep. Don't anchor too close inshore if you do stay overnight, in case of a shifting wind or a *vent solaire*.

ILE AUX MOUTONS

This small uninhabited island, with its unmanned lighthouse, lies five miles southeast of the mouth of the Bénodet River. In quiet settled weather you can anchor in the bay on the southeast side of Ile aux Moutons, which offers fair shelter from winds with any north in them. Approach from the southeast, having first reached a position close north of Les Pourceaux N Cardinal buoy. If coming from Bénodet or Concarneau, be sure to clear Pen ar Guernen, a ledge of drying rocks extending east from the anchorage for about a quarter of a mile.

BEG MEIL

This low headland at the southwest corner of Baie de la Forêt is easily identified by its prominent signal station. There are various anchorages off the west shore of the bay, to the north of Beg Meil, although local moorings occupy the best spots during the summer. Avoid Laouen Jardin, some rocky patches with barely a metre over them, 3–4 cables north by west from Beg Meil. Also avoid

RAZ DE SEIN TO BAIE DE LA FORÊT

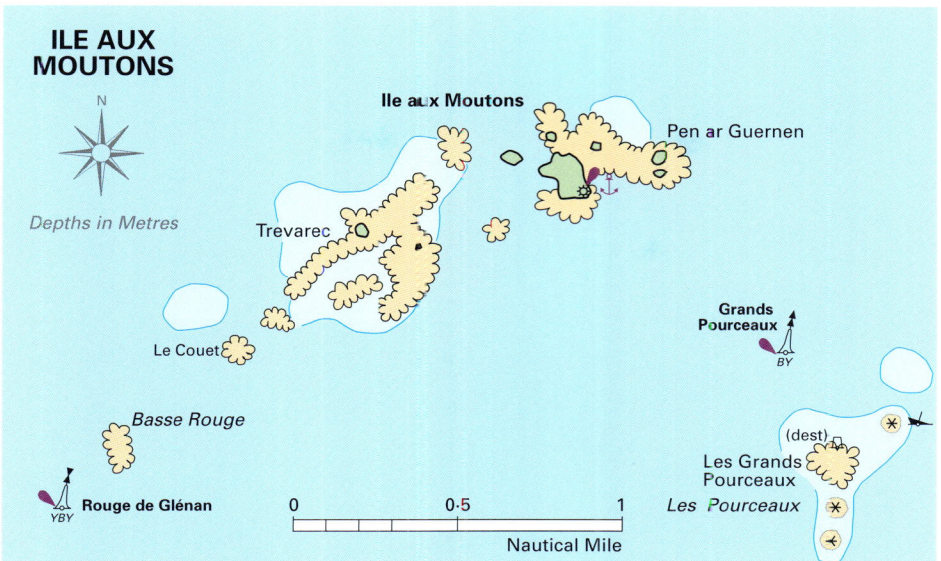

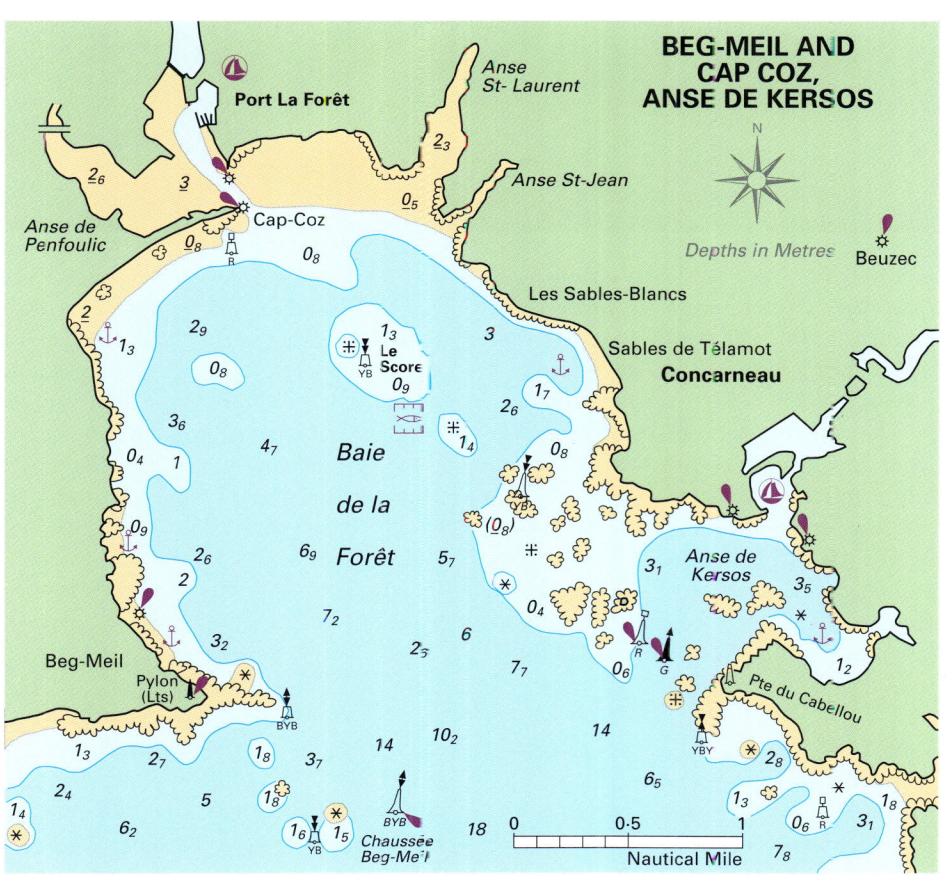

SECRET ANCHORAGES OF BRITTANY 189

Cap Coz and Port La Forêt

a couple of rocks awash at datum, less than two cables north by west from Laouen Jardin. The west side of the Baie de la Forêt is not well lit, so entering or leaving at night is not straightforward.

The various rocky shoals off Beg Meil itself are guarded by two beacon towers and an E Cardinal buoy. When approaching the bay from seaward or from the direction of Roches de Mousterlin and La Valeuse buoy, leave Linuen S Cardinal beacon tower and Chaussée de Beg Meil E Cardinal buoy well clear to port.

CAP COZ

There is an anchorage in the northwest corner of Baie de la Forêt, about 5-6 cables southwest of Cap Coz, the low narrow spit which forms the west side of the entrance to Rivière de la Forêt. Towards neaps, you can fetch up closer inshore towards the beach. This is a sheltered spot in fresh westerlies or northwesterlies, although the holding is mixed, with some rock and gravel patches amongst the sand. You'll see some waiting buoys for Port La Forêt about a quarter of a mile south of Cap Coz, close SSE of Les Ormeaux red beacon. The Cap Coz white sector and leading lights allow you to approach or leave this anchorage at night.

SABLES DE TELAMOT

This splendid stretch of beach opposite the village of Les Sables-Blanc lies just over a mile northwest of Concarneau harbour entrance, on the east side of Baie de la Forêt. Between Concarneau and Sables de Télamot, a wide area of reefs and shoals extends seawards for a

BAIE DE LA FORÊT

This magnificent sandy inlet a few miles east of the Anse de Bénodet provides an idyllic mini-cruising ground, where the low cliffs and white sandy beaches beckon a welcome. In the real summer weather for which this coast is noted, here are the ingredients for golden holiday memories, where the days blend into a timeless paradise of glittering blue water and just the right wind for wherever you choose to go.

If this sounds over-the-top, my own experiences of this area are consistently pleasurable, which I guess is unusual for cruising anywhere in northern Europe. When I think of Baie de la Forêt, Concarneau and Glénan I think of hot lazy sunshine day after day, with calm seas in the mornings to get you into gear slowly. I can picture yachts of all sizes motoring out into the bay with just their mainsails set, until the heat starts pulling in zephyrs of breeze off the sea. The French yachts set their genoas first and start ghosting across to Glénan or east towards the beckoning beaches off Trévignon and Raguénès The English usually motor a little longer, perhaps on a passage with a schedule to keep. Their white floppy sun-hats are almost classic logos for British yachts cruising this corner of France.

I can recall long idyllic days anchored off Beg-Meil or the Plage du Cap-Coz, when the very concept of work seemed light-years away and the vital decision of the moment was which wine to open for lunch. Long lazy cycles of swimming and sunbathing are part of childhood memories, but you can relive them again around Baie de la Forêt.

**Sailing towards
Port La Forêt**

good three-quarters of a mile. The final approach to the anchorage must therefore be from more or less due west, as you curve round to follow the deeper water in Baie de la Forêt clearly shown on Admiralty chart No. 3641. Sables de Télamot is a superb lunchtime anchorage on a quiet summer day, with excellent swimming over the sandy bottom. I've also stayed overnight here with the wind settled in the east or northeast.

ANSE DE KERSOS

This peaceful and attractive inlet lies on the east side of Baie de la Forêt, half a mile south of the entrance to Concarneau harbour. There are now plenty of moorings in this bay, but still room to anchor, especially near neaps when you can go well in past the moorings and fetch up off the beach.

La Croix light at the entrance to Concarneau

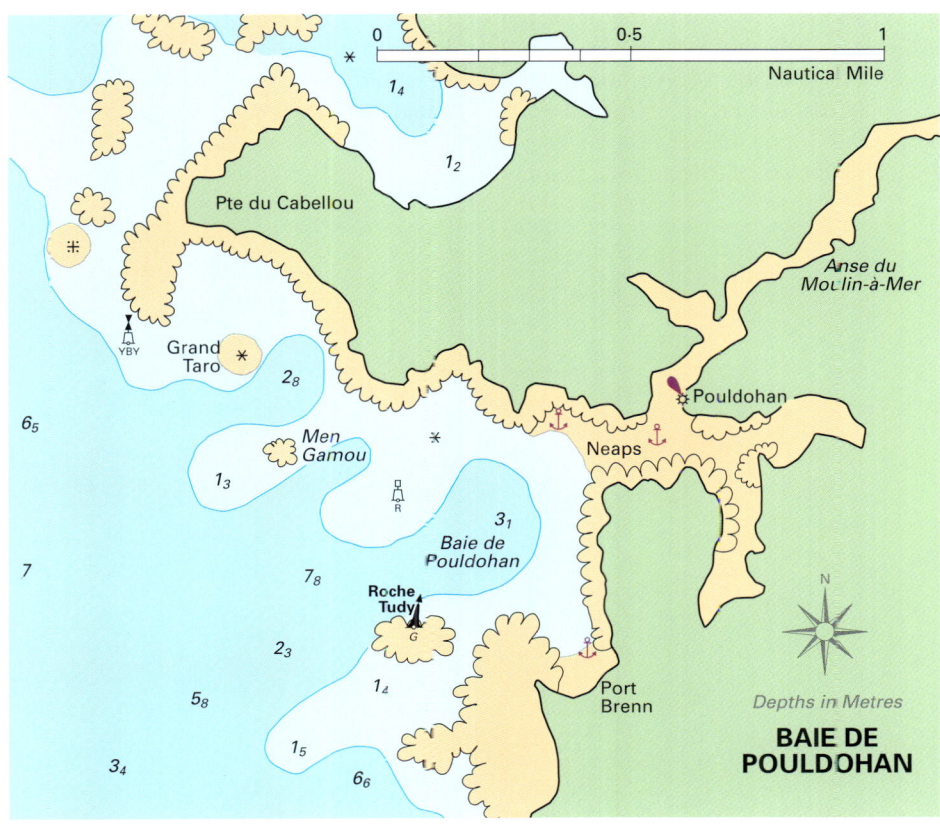

Leave Kersos green beacon tower a good cable to starboard when entering, since there are several drying rocks between this beacon and the west shore of Anse de Kersos. Work out your low water depth carefully before settling down for the night. The anchorage has pretty good holding and is well protected from between southwest through south to northeast.

BAIE DE POULDOHAN

About one and a half miles south of Concarneau, on the east side of Baie de la Forêt, the Baie de Pouldohan is bordered by a good many shoals and drying rocks. It is, however, interesting to visit in quiet or easterly weather using Admiralty chart No. 3641. Approach from a position three-quarters of a mile south of Pointe du Cabellou and then head ENE into the bay to pass midway between Karek Steir red spar beacon, left to port, and Roche Tudy green beacon tower, left to starboard. Anchor about a cable WSW of the inner green spar beacon, or a little further into the inlet if your draught and the tide allow. You can enter or leave Baie de Pouldohan at night using the green sector light on Pouldohan point.

Port Brenn is another possible anchorage in the southeast corner of Baie de Pouldohan, although this narrow cove is best entered near low water so that the extent of the various reefs on its west side can easily be seen.

KERCANIC

This wide sandy bight lies between Pointe de Trévignon and Pointe de Raguénès, but is the second bay east from Trévignon. In quiet weather or in winds with any settled north in them, you can anchor off the beach in the northwest corner of Kercanic bay, just outside the moorings. The roll is uncomfortable in even light onshore winds. Coming from the west, say from Bénodet or Concarneau, pass midway between Pointe de Trévignon and Men Du beacon tower; coming from the east, say from the Aven or Bélon rivers, most yachts pass between Ile Verte and Ile de Raguénès.

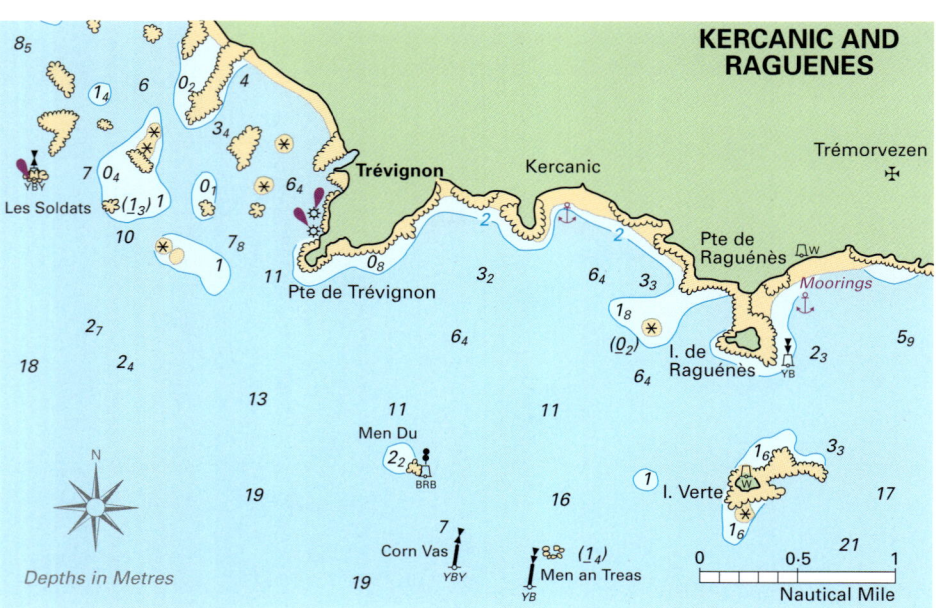

RAGUÉNÈS

Low-lying Ile de Raguénès lies close off Pointe de Raguénès, not quite three miles west along the coast from the mouth of the Aven and Bélon rivers. The east sides of the island and headland form a shallow bay in which plenty of small boats are moored during the season. There's an anchorage outside these moorings, accessible at all states of tide and reasonably sheltered from west through north to northeast. Sound carefully as you come in, because the water is shoal for quite a long way off the beach. Fair holding on a sandy bottom.

USEFUL ADMIRALTY CHARTS

No. 2819 Chausée de Sein to Penmarc'h
No. 2820 Penmarc'h to Pointe de Trévignon
No. 3640 Îles de Glénan
No. 3641 Loctudy to Concarneau

USEFUL SHOM CHARTS

No. 6648 Îles de Glénan – Partie Sud
No. 6650 Baie de la Forêt
No. 7249 Ports et Mouillages en Finistère Sud
No. 7423 Raz de Sein

Up river to Quimper

The mouth of the delightful Aven River opposite Port Manec'h

CHAPTER 6

AVEN RIVER TO PRESQU'ILE DE QUIBERON

The anchorages in this chapter are down in that congenial area of French Biscay between the Aven river and the Quiberon peninsula, including Ile de Groix. You could perhaps describe this cruising ground as 'near-Biscay' for British yachts, being accessible in a three-week cruise from most of our Channel harbours providing you aren't held up too much by weather on the way. This is a delightful coast, with sandy beaches, gentle tides and mainly straightforward pilotage.

You often experience a characteristic *vent solaire* in hot summer weather, which may set in from the west after a lunchtime calm and then blow onshore in the afternoon before veering through northwest to north by evening. The land breeze part of this cycle may start around midnight from the northeast, becoming quite brisk during the night until about breakfast time. The

AVEN RIVER TO PRESQU'ILE DE QUIBERON

The restful Aven River up at Pont-Aven

possibility of this sun wind has to be borne in mind when choosing overnight anchorages along this tempting looking coast.

The Aven and Bélon rivers flow into a common estuary some twelve miles east round the coast from Concarneau. On the west side of this estuary, just inland from Beg-ar-Vechen lighthouse, the village of Port Manec'h has a good landing quay and moorings just opposite in a tongue of deep water close west of the Aven bar. You can usually find room to anchor clear of the moorings.

Both the Aven and the Bélon rivers have a good many moorings, but there are still one or two spots where you can lie to your own ground tackle. The Aven is interesting to explore on the last couple of hours of the flood, and with a moderate draught you can get right up to the quay at Pont Aven.

The stretch of coast between the Aven river and the approaches to Lorient tends to be missed by visiting yachts, yet there are several small harbours and inlets which offer scope for anchoring overnight if the weather is quiet and settled. Brigneau, Merrien and Doëlan make up a fascinating trio not far east of the Aven and Bélon estuary. May or early June can be a good time to explore this area if you chance upon a friendly anticyclone; there won't be so many other yachts about and a *vent solaire* should be less likely to disturb your peace.

Three miles east of Doëlan is the rather tricky entrance to the shallow Rivière de Quimperlé. Although the bar and strong tides discourage visiting yachts, I have enjoyed some pleasant spells here when conditions were right. You certainly need either calm weather or light to moderate offshore winds, with

SECRET ANCHORAGES OF BRITTANY   199

Pont-Aven water mill

making tides midway between neaps and springs. Timing is all-important, to avoid the worst of the savage streams: you need to be entering or leaving the river in the hour before local high water. However, once you are safely in and anchored close under the west bank above Le Pouldu, there is no quieter spot along the South Brittany coast.

As you approach the three-mile wide strait between Ile de Groix and the entrance to Lorient, the atmosphere changes yet again. The outline of Groix is mysterious from a distance, higher than you might have expected, but gradually taking shape on the north side as a welcoming, steep-to coastline which provides a lee from any onshore swell. I have included three fair weather anchorages along this landward facing coast, and the island's main harbour – Port Tudy – lies towards the east end. However, the two most interesting anchorages off Ile de Groix – Port St Nicolas and Loc-Maria – are on the Atlantic side, vulnerable to swell but perfectly snug during a spell of northerlies.

The approaches to Lorient are well buoyed and dredged, as befits a large commercial and fishing harbour, but the big-ship channel can easily funnel yachts straight up to one of the Lorient marinas and divert attention from several possible anchorages along the mainland coast opposite Ile de Groix. Keroc'h, the Anse de Stole, Larmor, Locmalo and the Anse de Goêrem, are all useful natural havens off the normal cruising track, where you can enjoy a quiet and economical night. There are one or two

Ile de Groix crêperie

anchorages in Lorient harbour itself, but most of the areas free from commercial or naval activity are taken up with local moorings. If you are looking for seclusion and rural surroundings, the best bet is the Blavet River, which joins the east side of the harbour just before you reach the spur leading up to the Port de Commerce yacht harbour.

From Lorient entrance, once you have cleared the outer buoys of the Passe du Sud, low sand dunes curve gradually southeast to the mouth of the Etel river and then south towards the Quiberon peninsula. Although the first stretch is not suitable for anchoring, I've included a few coastal niches south of Etel, off the west side of the peninsula. Any significant swell rules out these anchorages, especially overnight, but during a settled spell of easterlies, even if the winds are fresh, you can find some good shelter in the locations I have indicated.

Aven river

AVEN AND BELON RIVERS

These picturesque rivers share a common estuary and are well covered by the pilot books, but each year it seems more difficult to find room to anchor. You can usually fetch up off Port Manec'h, at the mouth of the Aven just outside the bar,

Lorient harbour entrance

AVEN RIVER TO PRESQU'ILE DE QUIBERON

or off Port l'Hermite, a small inlet a little way into the Aven on the east side.

The picturesque Bélon river should be entered well above half-flood, when there's plenty of water over the bar at its mouth. Coming in, you need to keep close to the south shore of the river, leaving the green beacon off Pointe de Kerhermen about 100 metres to starboard and then heading just east of north to leave Pointe de Kerfany a similar distance to starboard. This line passes over a patch of rocks awash at LAT, but leaves the shallowest part of the bar to port. Thereafter, the channel trends over to the north shore before curving back to the south to follow the outside of a bend. There are visitors' moorings a quarter of a mile below Bélon quay off the north side of the river, but you can still swing to your own anchor just downstream from this trot.

BRIGNEAU

This small drying harbour, three miles southeast of the Aven river mouth, is rather susceptible to swell for taking the ground safely. In settled offshore weather, however, you can anchor and stay afloat at the entrance, a little way SSE of the outer pierhead. Approach from the SSE after half-flood, having first made a position to the east of the unlit RW fairway buoy moored about three-quarters of a mile south of the entrance. A S Cardinal beacon marks an isolated rock off the west side of the entrance and should be left two cables to port on the way in. A rock awash at datum, off the east side of the entrance, can be avoided by keeping the outer pierhead bearing 331°. You can enter or leave at night using the white sector of the Brigneau pierhead light.

Port Manec'h anchorage from eastward

MERRIEN

A picturesque creek and small harbour, not quite a mile east of Brigneau. The entrance is partly protected by a drying reef which extends half a mile seaward from the west side and whose extremity is marked by Le Cochon S Cardinal buoy. On the east side, a green spar beacon marks a patch of drying rocks two cables southwest of Pointe de Bali.

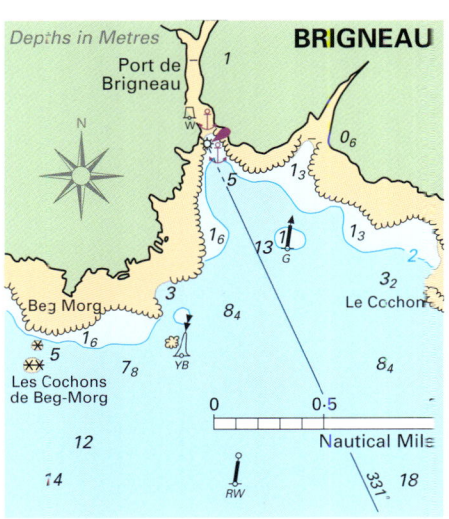

Bélon river landing

Port Manec'h

Approach from the south between Le Cochon and Pointe de Bali, bringing the white square lighthouse inside the creek in line with the middle of the entrance bearing 005°. It's best to enter the outer anchorage near low water, when the rocks either side are visible. Enter the river itself above half-flood, keeping to mid-stream. The pool opposite the seaward end of the east quay dries 0.8m to soft mud, but most yachts can stay afloat here at neaps.

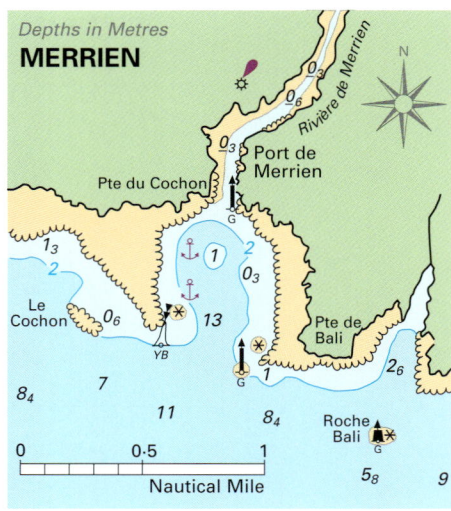

ARTISTS OF PONT-AVEN

The ever-changing light and spectacular seascapes around the coast of Brittany have always attracted painters from all over France and much of Europe. In the early 1880s Paul Gauguin gave up a lucrative career as a stockbroker in Paris to paint and in 1886 he left Paris, his wife and five children, to set up his easel in Pont-Aven. This picturesque South Brittany town at the navigable head of the Aven River is about halfway between Lorient and Quimper. Soon other artists had followed his lead and by 1888, together with Emile Bernard, Gauguin founded the Pont-Aven School. Within a very short time, some twenty artists were working in the town and its surroundings and the new style of 'symbolism' was developing.

Just before the Second World War the Scottish painter William Scott went to Pont-Aven to teach at the art school. He had mostly been a landscape painter until then but while he was at Pont-Aven his works turned more towards still life. He met both Emile Bernard and Maurice Denis at Pont-Aven and only left to return to Britain at the outbreak of hostilities when he joined the Royal Engineers.

In the town museum in Pont-Aven's Place de l'Hôtel de Ville, you won't see any of Gauguin's paintings – they have all been expensively collected elsewhere. However, other members of the Pont-Aven School are well represented, particularly Maurice Denis and Emile Jourdan. The temporary exhibitions in the museum often have important works on loan. You will also find here a fascinating photographic record retracing the history of Pont-Aven.

Many of the fourteen water mills that so often featured in paintings of the period are now converted into galleries themselves. There are seven waymarked routes around the Bois d'Amour, where those charming views that inspired so many famous painters are indicated so that present-day painters can also try their hands. You can walk out to the Trémalo Chapel where the wooden figure of Christ inspired Gauguin's *Yellow Christ*, or to the old church at Nizon where the calvary formed the basis of his *Green Christ*.

Pont-Aven lies where the narrow course of the Aven River starts to widen out into a navigable estuary. Not far downstream are the fine sandy beaches around Port Manec'h, a popular French seaside destination since the Belle Epoque. The Aven shares its mouth with the Bélon River, renowned throughout France for its distinctive and subtly flavoured oysters.

The charming old church at Pont-Aven

Local fishing boat
at Merrien

Merrien

DOËLAN

A small fishing port two miles east of Merrien. There is reasonable shelter inside, but the harbour is usually crowded with fishing boats and the swell can be nasty in onshore winds. In moderate offshore weather you can anchor outside, south of the pierhead. Approach Doëlan from the south, leaving Basse de la Croix red beacon one and a half cables to port and Le Four green buoy to starboard.

RIVIÈRE DE QUIMPERLÉ

This rather difficult river flows into the Anse de Pouldu, three miles east of Doëlan. The mouth has a bar and various shifting sandbanks, and the streams in the estuary are fierce at

AVEN RIVER TO PRESQU'ÎLE DE QUIBERON

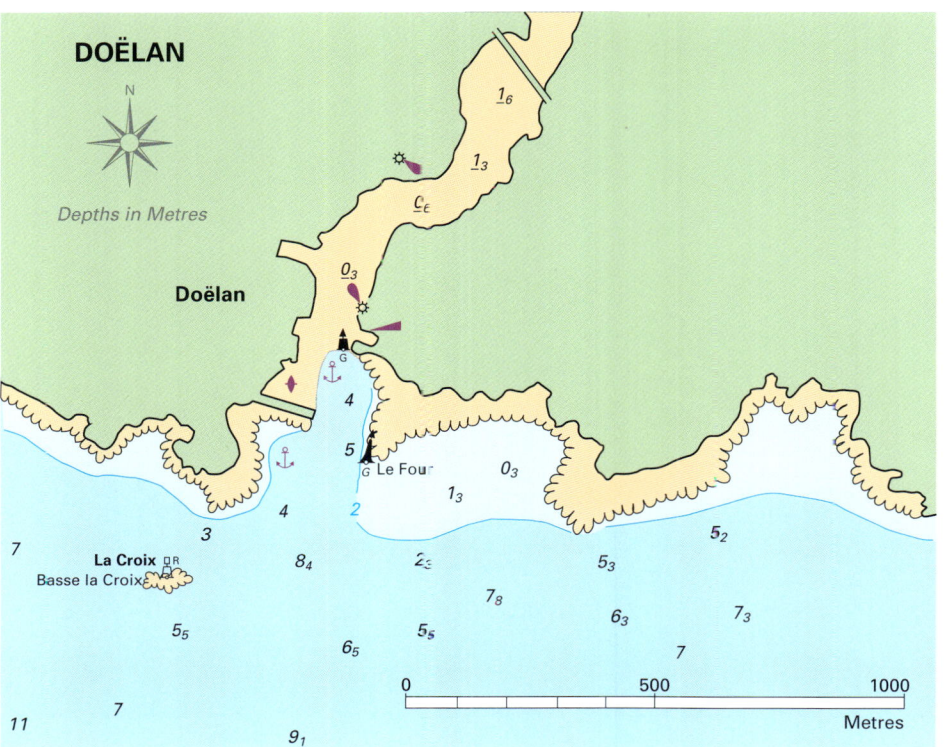

Doëlan

SECRET ANCHORAGES OF BRITTANY

OYSTERS OF THE BÉLON RIVER

The Bélon river is renowned for its oysters, cultivated in the shallow higher reaches beyond the moorings. Right on the quayside at Bélon, Chez Jacky and Les Huîtrières du Château de Bélon are popular haunts for oyster enthusiasts. The flat Bélon oysters are delicate-looking molluscs and the sandy-coloured flesh has a velvety texture and lingering taste, almost a fragrance, of hazelnut as it slips down. This is the distinctive *goût de noisette* for which Bélon oysters are renowned.

Les Huîtrières du Château de Bélon were established in 1864 by two pioneers in Brittany oyster cultivation, Hippolyte de Mauduit and his brother-in-law Auguste de Solminihac. It's said that the particular blend of freshwater and seawater in the Bélon river gives these highly prized oysters their special flavour. Les Huîtrières is now run by François de Solminihac, a direct descendent of the innovative Auguste.

Chez Jacky has been a favourite with visiting yachtsmen for many years, specialising in *fruits de mer* and particularly local Bélon oysters. This notable riverside bistro is run with panache by Sylviane, widow of Jacky Noblet who started the business. Along with the renowned restaurant, with its wonderful views across the river, Chez Jacky also produces oysters and *fruits de mer* which are sent all over France to both commercial customers and households. You can phone up and order a tasty lobster, a couple of dozen Bélon oysters or a *plateau de fruits de mer* to be delivered to the door either for that special occasion or simply for a 'routine' Sunday lunch.

Bélon River

springs. Entry should only be attempted in quiet offshore weather, about an hour before HW, and preferably halfway between neaps and springs – at neaps there is too little depth and at springs the tides are too strong for safety.

Identify the west headland of the entrance, on which stands a white house with a distinctive round tower. Approach this headland from a little west of south and make for Men-Du red beacon tower at the foot of the cliffs. Leave this tower and the red spar beacon beyond it about 50 metres to port, and then curve to port until the red spar is more or less in transit astern with Men-Du.

From this line follow the curve of the west shore into the river, leaving another red spar beacon about 50 metres to port. From this second red spar steer 045° through the entrance narrows, a course that should leave some submerged wooden posts at the end of a low sandy promontory on the east shore to starboard. Once you are right in the middle of the narrows, turn north to follow the west bank about 150 metres off, steering to leave the next headland with its prominent hotel about 100 metres to port. Opposite on the east bank you'll see a small *port de plaisance*, with a green spar beacon marking the end of the marina breakwater.

Just beyond the hotel you pass Port du Pouldu on the west bank, with its landing slips and drying hard. Now follow the line of moorings close past the west shore and anchor beyond these

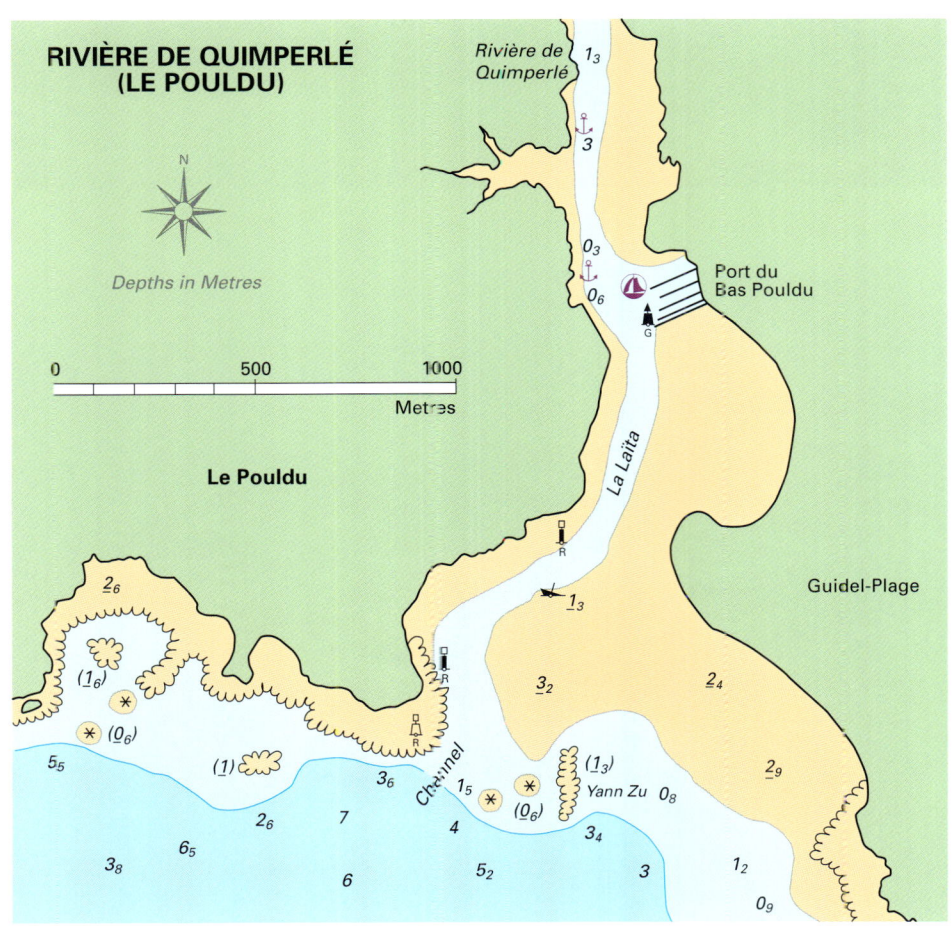

Le Pouldu
Keroc'h

moorings opposite the mouth of a shallow inlet known as Anse de Stervilin. This narrow sheltered pool has about two metres minimum depth at datum and you'll be pretty likely to have it to yourself.

Don't try to enter or leave the estuary in onshore winds or when there is any significant swell running.

KEROC'H

A small harbour on the mainland opposite Ile de Groix, half a mile northwest of Pointe du Talut. There is room to anchor clear of the local moorings and partly protected by the breakwater. Les Soeurs rocks lie to the west of the harbour and are unmarked. The safest approach is from due south, steering to leave Les Loups W Cardinal beacon tower 50 metres to starboard. Once past Les Loups, come to starboard for the breakwater head. Keroc'h is rather exposed from between west and northwest, but is otherwise well sheltered.

LOMENER

A useful passage anchorage just over a mile east of Pointe du Talut and one and a half miles NNW of Loqueltas S Cardinal buoy, which is the outer buoy for the Passe de l'Ouest into Lorient. Approach Lomener from due south above half-tide, steering to leave Grasu S Cardinal beacon tower a quarter of a mile to the east. The charted leading line is difficult to see because the ubiquitous 'church spire' is masked by trees, but the forward mark – Lomener white light tower with an orange top – is easy to identify. Keep this tower bearing more or less due north true to leave Les Trois Pierres (0.9m at datum) clear to port opposite Grasu and then leave a S Cardinal spar beacon (standing about a

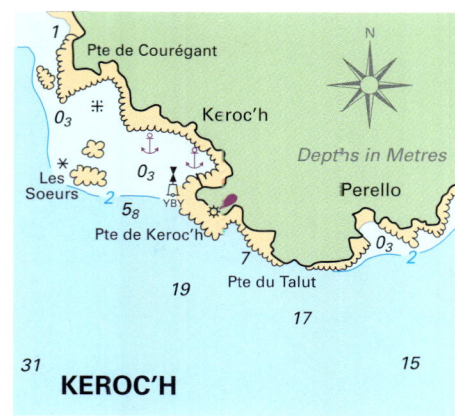

cable south of the breakwater head) a short 200 metres to port.

There are plenty of local moorings behind the breakwater, but you can anchor just outside them, east of the breakwater head with the light tower still bearing due north or a little more. Sometimes you can carry on north into the Anse de Stole and use a vacant mooring there. Anse de Stole is bordered by wide rocky ledges, so keep near the middle of the bay. There are shops and restaurants at Lomener, and water at the quay.

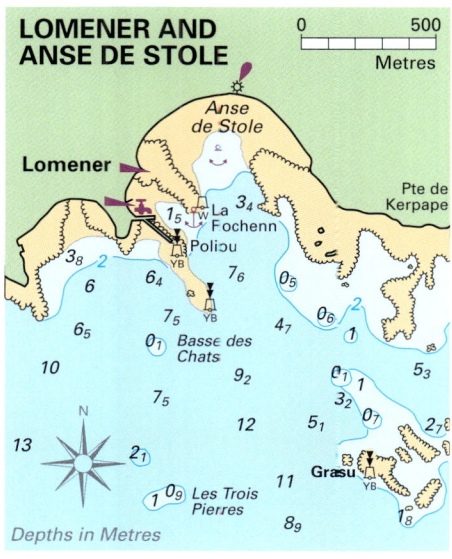

Anse de Stole

ILE DE GROIX

ILE DE GROIX

Only a dozen miles southeast of the Aven River, the intriguing profile of Ile de Groix often beckons. Only four miles from end to end and barely one and a half miles across, Groix nevertheless has the substantial feel of a real island when you get up close, with its neat looking villages and regular mainland ferry. Port Tudy, the island's only snug harbour, is on the north side, facing the approaches to Lorient. The best chart for all the Ile de Groix anchorages is the large-scale French SHOM No. 7139.

Port Tudy Roads The attractive harbour at Port Tudy, on the north coast of Groix, is popular with visiting yachts and well covered by the pilot books. However, it's worth remembering that you can anchor to the northwest of the outer breakwater in quiet weather or in moderate southerlies. This can be an agreeable option if Port Tudy is particularly overcrowded.

Beg-ar-Vir In quiet or southerly weather you can anchor off the beach at Beg-ar-Vir, on the north coast of Groix

Port Tudy waterfront on Ile de Groix

Pen Men, Ile de Groix

about one and a quarter miles west of Port Tudy and a similar distance east of Pen Men lighthouse. It's important to keep between the two ledges of drying rocks either side of the anchorage, so come in from due north, preferably near low water.

Port St Nicolas The southwest coast of Groix is exposed to the Atlantic swell and sometimes looks rather inhospitable, but when high pressure reigns the sea is often glassy calm in high summer and it's perfectly safe to explore round the island. There's a fascinating inlet between the rugged cliffs about one and a quarter miles southeast of Pen Men lighthouse. Make the final approach from the southwest, leaving Pointe St Nicolas to starboard and heading for the middle of the inlet.

Fetch up in about two metres at datum, but watch out for a ledge of rocks that extends into the head of the cove from the north side. The bottom is sandy with plenty of rock and thick weed, so a good Fisherman's anchor (buoyed) is the best bet. Port St Nicolas is sheltered from between northwest through north to east, but clear out at the onset of any onshore wind or southwesterly swell.

Loc Maria An unspoilt natural harbour in the southeast corner of Groix, about three-quarters of a mile WNW of Pointe des Chats and sheltered from winds with any north in them. The

AVEN RIVER TO PRESQU'ILE DE QUIBERON

Loc Maria
Pointe de la Croix

SECRET ANCHORAGES OF BRITTANY ⚓ 215

LE VENT SOLAIRE

Those who cruise frequently on the Biscay coast in high summer will be familiar with the *vent solaire*. This slightly unpredictable thermal wind is often attributed a rather devilish character by anyone who has been caught unawares and found their snug overnight anchorage off one of the offshore islands transformed unexpectedly into a lee shore. In fact the *vent solaire* is no more than the classic land breeze we are all familiar with in hot weather, particularly in late summer when the sea has warmed up well.

Sea breezes get going towards the middle of a summer day when the land is heated by the sun, the warm air immediately above the land starts to rise and air is drawn in off the sea to take its place. At night, if the land cools down below the coastal sea temperature, the reverse process takes place, relatively warm air rises over inshore waters and air is drawn out from the land to take its place. This is the land breeze, which can get up in the chilly early hours and blow offshore until the sun gets well up and starts reversing the process. Cruising yachts often make good use of coastal land breezes overnight, allowing them to reach along nicely under a weather shore. But such land breezes can become inconvenient if you happen to be anchored off the landward side of an island not very far offshore.

In this case the 'land' breeze off the mainland coast will become a 'sea' breeze in your hitherto sheltered anchorage. This phenomenon is fairly common in summer on the landward sides of the Biscay islands such as Ile de Groix, Belle Ile, Houat and Hoëdic. Such *vents solaires* are indeed rather fickle and don't always spring up when you expect them to, even in very warm weather. They are more likely to get going on clear nights when the mainland is able to cool down more quickly than if it were covered with cloud.

French old gaffer at sea off Houat

harbour itself dries and plenty of small local boats lie on moorings here during the summer, but there's room to anchor and stay afloat just outside the harbour area, with the east pierhead bearing 060° about 250 metres distant.

Coming from the east, or arriving east-about from the north coast, it's important to clear Les Chats by a safe margin – this extensive area of drying rocks reaches out for nearly a mile south from Pointe des Chats. You don't need to go right round Basse des Chats S Cardinal buoy, but you must certainly pass within a quarter mile of it before making any westing. Also bear in mind that the streams are quite strong at springs off this part of the island. Because the leading marks for Loc Maria are rather tricky to identify clearly from offshore, it's best to enter within a couple of hours of high water, when it's not so critical to stay on the charted approach line.

From an offing position one and a half miles SSW of Pointe des Chats lighthouse, head north towards the green beacon tower that marks the east side of the entrance to Loc Maria bay. An isolated rock drying 0.4m lurks just over 150 metres south of this tower, but is safely covered within two hours of high. Leave this beacon tower 100 metres to starboard and then head NNW to pass between the western pair of red and green spar beacons (don't use the eastern pair 300 metres north of the beacon tower).

From this western gateway steer towards a white beacon on the shore on the west side of the harbour bay and anchor when a N Cardinal spar beacon to starboard draws just abaft the beam. At neaps you can edge a little further northeast from this position, towards the head of the east jetty. Clear out quickly if the forecast hints at onshore winds or if any significant swell starts filtering in.

Don't attempt to enter or leave Loc Maria at night.

Pointe de la Croix In quiet settled weather, you can anchor off the beach close north of Pointe de la Croix, at the northeast corner of Ile de Groix. This attractive spot is usually well protected from southwesterly swell. If you opt to stay overnight and the wind shifts, it's straightforward to move round to the snug harbour at Port Tudy.

LARMOR

A rather open anchorage on the west side of Lorient harbour entrance, but well protected from between west and north. As you are approaching the narrows at St Louis, turn to the northwest at l'Ecrevisse red buoy and make for Larmor breakwater head. The final approach leaves two red spar beacons to port. Fetch up clear of the local moorings. Good shops and restaurants ashore. If the wind should shift overnight, it's easy enough to reach the shelter of Lorient harbour. Use Admiralty chart No. 304 for getting into Larmor.

Lorient harbour

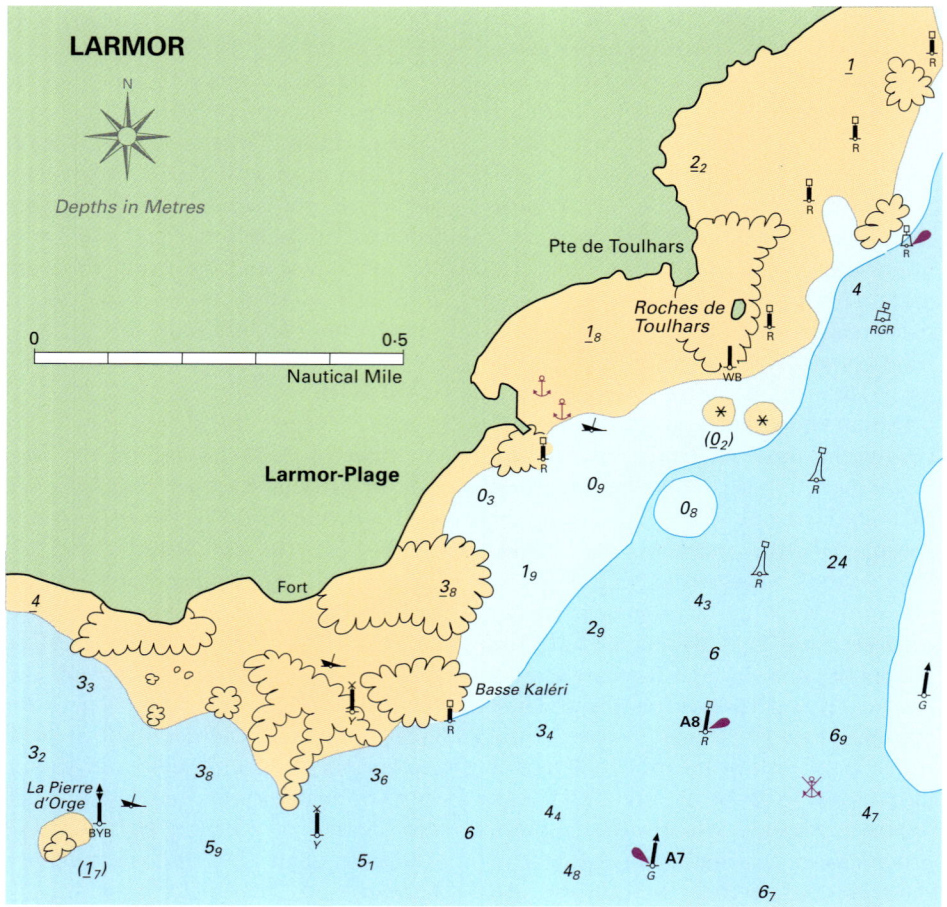

ANSE DE GOEREM

A useful anchorage in easterlies, off the west side of the Gâvres peninsula. From the buoyed approaches to Lorient, make good due east true from midway between Le Goëland and Basse de la Paix green buoys and fetch up about a cable south of Le Pesquerez green spar beacon, or further inshore at neaps. Use Admiralty chart No. 304.

BAIE DE LOCMALO

This pleasant sheltered inlet opens out to the east of Lorient harbour entrance, close south of the St Louis peninsula. Most of the bay dries on a good spring except for a shallow gully, but at neap tides (or at least midway between springs and neaps) most boats should be able to stay afloat at low water. Ideally, you should enter or leave Baie de Locmalo within a couple of hours of high water.

The most straightforward approach is to turn off the Lorient entrance channel three cables south of the St Louis narrows and pass north of La Potée de Beurre green beacon tower. Steer ESE towards the north edge of Ile aux Souris at first and then nudge to port to leave a green light structure and then the island itself to starboard. Now follow the line of the north shore to pass between Petit Belorc'h red beacon tower

SUBMARINES AT LORIENT

The Battle of the Atlantic between the German navy and Allied convoys lasted practically the whole of the Second World War. Britain depended crucially on her American supply lines and Hitler reasoned quite rightly that if Atlantic shipping routes could be cut or severely disrupted, Britain would eventually be starved into a state of weakness. Germany's deadly weapons in this cruel warfare were Admiral Karl Dönitz's growing fleets of U-boats, their skippers and increasingly young crews much fêted by Hitler as the naval *crème de la crème*. U-boats were more than just highly effective fighting machines. Their sinister invisibility worked on the morale of all seamen involved in transporting supplies across an ocean that was already hazardous enough.

Hitler established five secure U-boat bases on the French Atlantic coast at Brest, Lorient, St Nazaire, La Pallice (near La Rochelle) and Bordeaux. The Lorient base is particularly intriguing because you can actually visit this rather eerie place by joining a guided tour run by the local tourist office. Lorient's narrow entrance is guarded on its east side by the Port Louis citadel, whose sheer walls look right for some boiling oil. On the west side, amidst the seaside clutter of Larmor-Plage, is the rather garish but comfortable Château Kernével where Dönitz established his Lorient headquarters. As the fine natural harbour opens out, you see the stark row of Lorient's submarine pens as a backdrop to the massed masts of Kernével marina. This dramatic effect is oddly attractive seeming to affirm the triumph of peaceable leisure over ravaging war. Although a visit to the U-boat pens may sound a gloomy prospect, to see the scale of this engineering at close quarters is a remarkable experience that brings home Hitler's staggering grandiose intent.

Wandering inside these cavernous bunkers that still contain original German lifting gear for loading torpedoes, you can't help reflecting on the precarious pivot of events that could have swung the outcome of 1939-45 the other way. Throughout the Battle of the Atlantic, Dönitz kept closely in touch with his commanders and crews from his Lorient base, inspiring them to extraordinary feats of endurance in chilling conditions. Totally committed to Hitler's grand project, Dönitz became Grossadmiral in 1943 and later C-in-C of the German navy. After Hitler's suicide in 1945, Doenitz was briefly the last Führer of Germany.

Lorient submarine pens

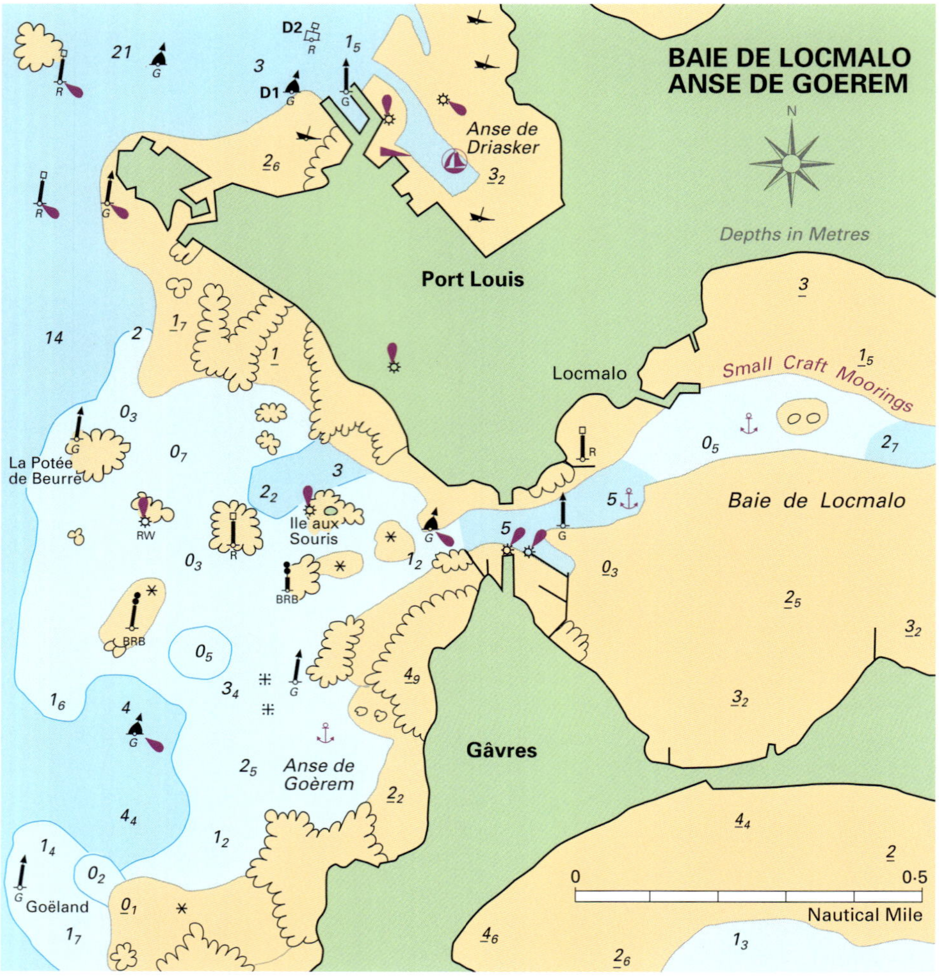

and Grand Belorc'h green. Anchor clear of the local moorings, working out your low water depth carefully. At dead neaps most boats will stay afloat in the tongue of water ESE of the Locmalo breakwater. The French SHOM chart No. 7140 is best.

LE BLAVET RIVER

This rather forgotten but attractive river joins the northeast corner of Lorient harbour just below the west turn that leads up to the city marinas. Around neap tides, using the French SHOM chart No. 7140, boats of moderate draught will have a choice of peacefully remote anchorages in the first couple of miles of the Blavet, where the channel is marked by buoys and beacons. Set off upstream just after half-flood and watch the echo-sounder carefully. If you decide to anchor overnight, work out your low water depth carefully.

ERDEVEN

In settled easterly weather, there are two secluded spots where you can anchor off the dunes to the southeast of Pointe d'Erdeven, a couple of miles south along the coast from the Etel river. The first is opposite a narrow inlet, just half a mile southeast of Pointe d'Erdeven. It's easiest

AVEN RIVER TO PRESQU'ILE DE QUIBERON

Blavet river

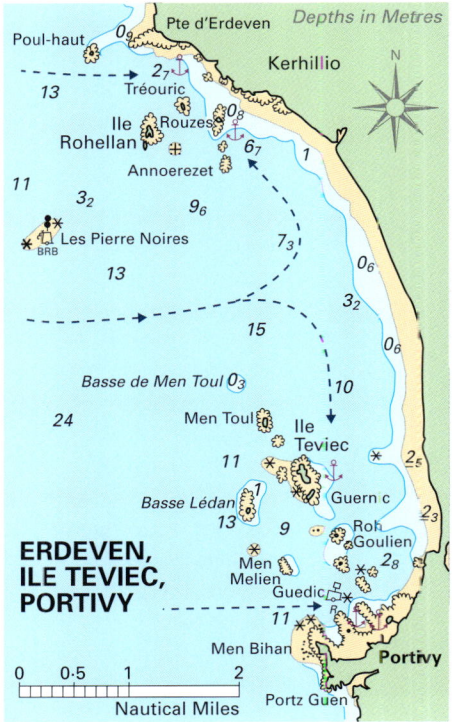

to approach near low water, when the various rocky patches surrounding the anchorage can clearly be seen. Close the coast from one and a half miles offshore, making good due east to pass midway between Poul-haut rock (3m high), left to port, and Ile Rohellan, left to starboard. Now aim to pass south of the rocky ledges bordering Pointe d'Erdeven and north of Tréouric rock, which lies just over a quarter of a mile northeast of Rohellan. Fetch up in two metres at datum, about two cables north by east of Treouric.

The second anchorage is a mile further southeast, in behind some more drying rocks that lie up to three-quarters of a mile southeast of Tréouric. Approach this anchorage from further south, leaving Les Pierres Noires beacon tower half a mile to the north and heading east to close the beach to within three-quarters of a mile. Now turn north

SECRET ANCHORAGES OF BRITTANY 221

SEMAPHORE PILOTS AT ETEL

The Etel River meets the sea on a low coast of sand dunes, about seven miles southeast of Lorient entrance. While much of the lower river is quite deep, the mouth is precariously shallow, with a notorious bar forming a dangerous hump between two drying spits jutting out from each side of the entrance like invisible jetties. In fact, this whole stretch of coast is shallow for a good third of a mile offshore, so you can easily imagine how the southwest facing mouth of the Etel becomes a seething mass of breakers with an Atlantic swell rolling in, especially during the lower half of the tide. The Etel River is also fast-flowing and a good run of spring ebb stirs things up even further.

The position and depths of the bar and spits have always been prone to change after onshore gales, when you could only discover the current shape of the bar and trend of the channel on the next low water. In the days of sail, the local tunny fishermen devised a curious yet highly practical pilotage system using a traditional semaphore signalling station – Le Mât Fenoux – with two large mechanical arms operated by levers and pulleys. The station still stands amidst the dunes on the north side of the entrance, conspicuous by the white gable end of its high-pitched Breton roof. The semaphore has been run since 1980 by the redoubtable Madame Josiane Péné, who took over the levers after her husband died. Madame Josiane is famously the only woman pilot in France.

The semaphore system was simple but effective. When a boat arrived off the entrance, she would heave to at least half a mile offshore and hoist an ensign at the masthead to show that she wished to enter the river. The pilot ashore would waggle the semaphore arms to show that the boat had been spotted and was now under expert guidance. Then the pilot would simply con the boat in using the semaphore arms – right-hand arm out: steer to starboard, left-hand arm out: steer to port, both arms straight up: steady as you go.

The traditional semaphore method can still be used by cruising boats and the station continues as a cherished French maritime institution in Madame Josiane's hands. The initial contact is now usually made by VHF on Channel 13 rather than by hoisting an ensign, and indeed Josiane may actually direct you in by radio. However, on high days and holidays the semaphore arms waggle into life and you can follow their reliable directions across the bar with a sense of history and, when the weather is edgy, a real seaman's gratitude.

Etel semaphore station

Ile de Groix ferry off the Quiberon peninsula

along the coast, edging inside Annoerezet rock and fetching up between this rock and the shore. There is slightly less water in this second anchorage, so it's best visited near neaps.

You need to be sure of the weather before staying overnight at either anchorage, because this coast is not safe to navigate at night. Refer to Admiralty chart No. 2822 or, preferably, the French SHOM chart No. 7032.

ILE TEVIEC

This tiny island lies about a mile offshore, some four miles south by east from Pointe d'Erdeven and one and a half miles north of Presqu'île de Quiberon. In quiet weather or settled easterly winds, there's an anchorage off its east side, with the island and various off-lying reefs affording some protection from any slight westerly swell. Approach

Ile Teviec

Portivy entrance

Portivy harbour

the coast from further north, passing half to three-quarters of a mile south of Les Pierres Noires beacon tower and making good due east true to close the beach to three-quarters of a mile. Now turn due south to follow the coast three-quarters of a mile off, fetching up no closer than two cables east of Ile Teviec. Bear in mind that this stretch of coast is unlit, so the anchorage is not safe to leave at night should the weather conditions change.

PORTIVY

A small fishing harbour right on the northwest corner of Presqu'île de Quiberon, Portivy offers a pleasant anchorage in settled easterlies, so long as there is no significant Atlantic swell running. From a position half a mile WNW of the distinctive promontory of Beg en Aud, identify Guédic red buoy and approach it from due west, so as to leave the rocks off the tip of Beg en Aud a good two cables to the south. Leave the buoy close to port and then turn to starboard towards a red beacon standing about four cables southeast of Guédic, keeping midway between the two rocky ledges that enclose the channel to the harbour. At springs you should anchor just before reaching this red beacon, but at neaps you can leave it close to port and edge a little further in towards the breakwater. It's not safe to enter or leave Portivy at night.

USEFUL ADMIRALTY CHARTS

No. 304 Lorient Harbour (including the Blavet river)
No. 2821 Ile de Penfret to Plateau des Birvideaux
No. 2822 Ile de Groix to Belle Ile

USEFUL SHOM CHARTS

No. 7031 Ile de Penfret au Plateau des Birvideaux
No. 7032 Ile de Groix à Belle Ile
No. 7139 Ile de Groix
No. 7140 Passes et Rade de Lorient
No. 7141 Baie de Quiberon

Locals pottering off the Aven River

Anchored in the timeless Gulf of Morbihan

CHAPTER 7

BELLE ILE TO THE GULF OF MORBIHAN

Quiberon Bay is one of those stretches of water, like Plymouth Sound or Spithead, which has an unmistakable frisson of being steeped in history. You can sense the past almost as soon as the land is hull-up – when you first spot the hard outline of Belle Ile and then pick up the low Quiberon peninsula, from which a string of rocks and shoals extends for several miles to the southeast.

On then to Belle Ile, whose rather forbidding profile at a distance becomes only slightly less harsh as you draw near. This is the largest of the Brittany islands, nine miles long and a good four across at its widest. The main harbour, Le Palais, lies about halfway along the landward side, with a smaller harbour, Sauzon, near the northern tip. This landward coast is the least gaunt, and there are some pleasant anchorages,

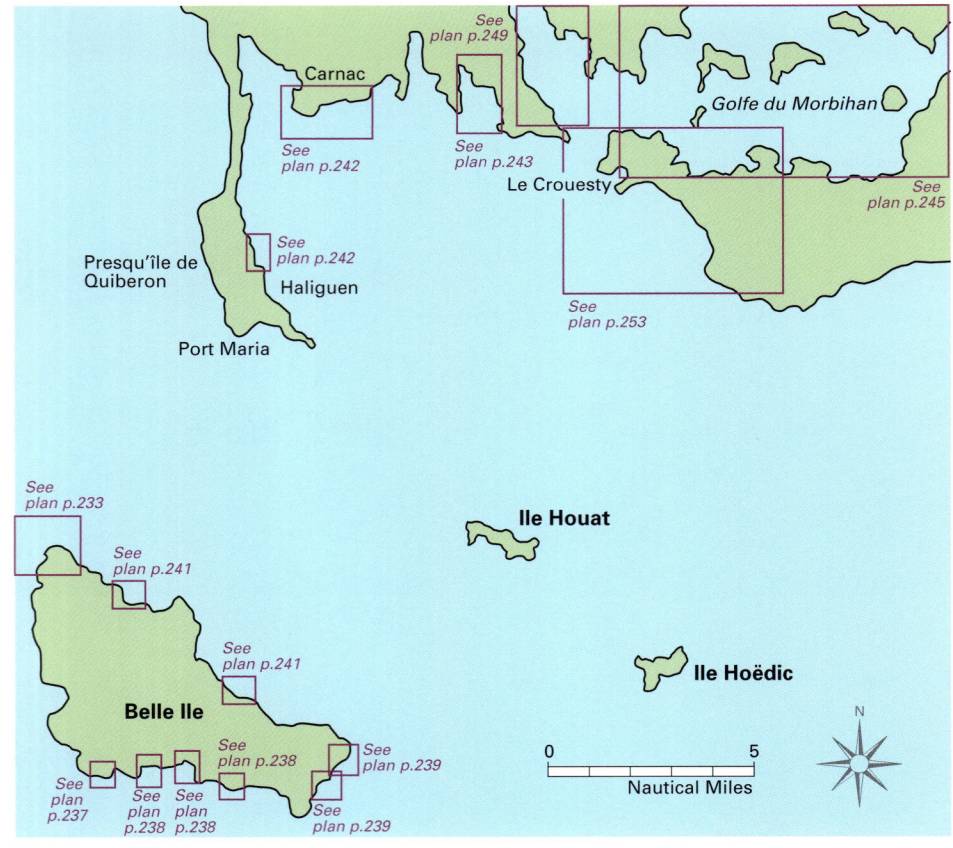

228 SECRET ANCHORAGES OF BRITTANY

Ster-Vraz and Ster-Wenn, Belle Ile

rather reminiscent of the West Country, in the bays southeast of Le Palais. I use these handy anchorages as overnight passage stops.

On the Atlantic side, the cliffs are steeper, well weathered and heavily indented, with few signs of habitation to relieve the rather intimidating façade. Some of the deeper indentations have become pronounced inlets over the centuries, and these can provide dramatic anchorages when, during a calm spell or a period of northeasterlies, the Biscay swell is temporarily at rest.

Belle Ile was seized by the English in 1572 and later exchanged for Menorca under the Treaty of Paris in 1763. In 1759, with great naval panache, Admiral Hawke led his fleet in amongst the rocks and shoals of Quiberon Bay during a winter gale to win the famous battle against a French invasion fleet under the Marquis de Conflans. It's partly all these seafaring ghosts that make you want to seek out natural anchorages hereabouts. These waters have such a reputation for adventure under sail that it seems decadent in the extreme to lounge about in large marinas at Port Haliguen, La Trinité or Crouesty. Such is the layout of the bay and its off-lying islands that you can usually find a snug retreat somewhere, and if you do get a spell of really nasty weather there are dozens of anchorages in the sheltered confines of the Morbihan.

The west side of Quiberon Bay is bounded by the Presqu'île de Quiberon and its long sandy isthmus, a popular area for the French on holiday. There are one or two anchorages off the inshore coast of the Presqu'île, to the north of Port Haliguen. The northwest corner of Quiberon Bay has a wide shallow inlet, between the isthmus and Pointe St Colomban, before the coast curves round to the east past Carnac-Plage, La Trinité, Anse de St Philibert and the

Gulf of Morbihan

Entrance to the Gulf of Morbihan

entrance to the Gulf of Morbihan. I have included several anchorages along this stretch before taking the tide through the narrow gut into the timeless Gulf of Morbihan.

Here you'll find some 50 square miles of sparkling tidal water almost completely enclosed by the two long peninsulas of Rhuys and Locmariaquer. Even the name of this French cruising paradise conveys the essence of coastal Brittany. *Mor Bihan* – 'a small sea' in the Breton language – has a special cadence which, if you know the place a little, recalls those haunting estuary vistas and the salty zest of the Atlantic spreading inland to create a perfect puzzle of tidal channels, tiny islands and secret shallow bays. To spend a long summer here, pottering between anchorages and lingering lunches, would be to experience the most idyllic features of Brittany in a single compact cruising ground.

The entrance gap, less than half a mile wide, lies in the northeast corner of Quiberon Bay, behind a natural cordon of drying shoals and islets. The approach is not difficult and the shoals are well marked, but the key to coming or going is timing the tide. And you need to know what to expect.

With such a vast expanse of water squeezing in or out, you can experience up to nine knots in the entrance at top springs, with dramatic tide rips, eddies and overfalls all around. It's easy to be catapulted into the Gulf before you are ready to face the pilotage decisions presented by a bewildering array of islands and enigmatic channels, their perspective shifting rapidly as you are set sideways without really noticing. For a

Up the Auray river

first visit, choose a period of neaps, so that you can get the measure of things in more moderate streams. The area around the entrance is the most active. Once you get your bearings and start meandering inland, things quieten down except in certain narrow straits and constricted corners.

Sluicing in past Port Navalo and intent on avoiding Grand Mouton green beacon, your immediate choice is to bear to port for the Auray River, or come to starboard round Grand Mouton towards the main body of the gulf. The Auray River is a tempting first bet and its seven miles of wooded banks, shallow creeks and oyster beds change little over the years. The streams aren't so strong as those between the islands of the gulf, and there aren't so many other boats about. The channel is wide enough above half-tide for working up under sail, at least as far as Le Rocher, although in the lower reaches you should stay in the main fairway round the east and north of Grand Harnic islet. Just opposite Grand Harnic, to starboard, is the entrance to the peaceful Anse de Badène.

The river begins to narrow above Pointe de Kerlevarech and there are various possible anchorages up as far as Le Bono. A good spot is three-quarters of a mile above Kerlevarech, off the east bank in a small bay opposite the shallow bight known as Port Espagnol. Less than a mile further upstream, the wooded banks edge close together before Le Rocher. The last stretch up to Auray is only navigable above half-tide, marked mainly by starboard-hand spar beacons. Auray is a charming old town and the St Goustan quarter, clustered around the quayside, is picturesque, with many of its timbered houses dating back to the 15th century.

Back down at the Morbihan entrance, the main channel leading northeast into the gulf leaves to port the

St Goustan, Auray river

southern ends of Ile Longue, Ile Gavrinis and Ile Berder. To starboard you pass Er Lanic islet with its green spar beacon, and then the north end of Ile de la Jument with an off-lying green conical buoy. The tide in this stretch can be fast and turbulent, but even under sail you'll be carried through safely just by keeping to the middle.

Parts of the Morbihan are shallow, with about half the charted area shown as drying at LAT, but there are plenty of deep channels between the numerous islands and enough anchorages to keep you busy for a whole summer. Although the tide can pour through the entrance and between some of the islands at a spectacular rate, there's usually no problem about piloting your way about, once you get used to ticking off landmarks rather quickly as they flash past. The trick is to hold your nerve.

The attractive town of Vannes lies in the northeast corner of the gulf, its sheltered locked basin reached by a winding channel above Conleau narrows. Vannes is a busy provincial centre and its old quarter is enclosed by 13th-century ramparts. The locked basin is a pleasant place to lie, handy for topping up with stores before you explore the anchorages of the gulf. Because Vannes is on a main railway line and also less than three hours drive from St Malo, it can make a convenient base to leave a boat or change crews.

Outside the Gulf of Morbihan, beyond the entrance to Le Crouesty marina, the coast trends southeastward for a couple of miles in a shallow bay between Petit Mont and Pointe du Grand Mont. I have included a couple of anchorages in this bay, which can be used overnight in northeasterlies or just

BELLE ILE TO GULF OF MORBIHAN

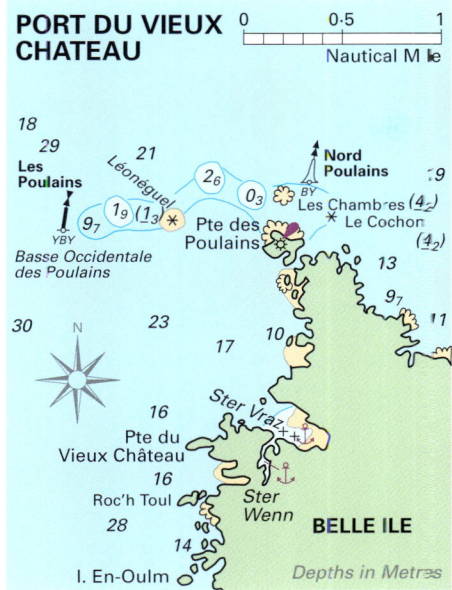

island's capture by the English in 1572 and ended with its exchange for Menorca under the 1763 Treaty of Paris. You can see the remnants of this tussle in Le Palais, where the port is overlooked by an imposing citadel fortified in the late 17th century by the ubiquitous Sébastien le Prestre de Vauban, who for 53 years was Louis XIV's chief military engineer. You can't cruise far around the French coast without coming across a Vauban fortification of some kind.

The outline of Belle Ile can look sombre even on a calm sunny day, but more so when breakers gleam at the foot of the cliffs. Yet despite its craggy aspect, pilotage around Belle Ile is fairly straightforward in quiet summer weather, with much of the coast steep-to and most of the granite visible. The tidal streams are generally moderate along the coasts, about one and a half knots at springs, but stronger close to the northwest and southeast ends of the island – Pointe des Poulains and Pointe de Kerdonis respectively Night navigation along the seaward coast is not recommended, although on the landward coast it's not difficult to leave

for a few hours in quiet weather to wait for a fair tide into the Morbihan.

BELLE ILE

Largest of the Brittany islands, Belle Ile has a distinctly chequered past, a focus for two centuries of wrangling between England and France that began with the

Ster-Wenn anchorage

SECRET ANCHORAGES OF BRITTANY 233

BATTLE OF QUIBERON BAY

To enter Quiberon Bay you have to thread the rocky shoals that straggle out from the Quiberon peninsula for nearly five miles. Anyone who has dodged through La Teignouse passage can imagine the fine seamanship involved in November 1759 when Admiral Edward Hawke led his fleet into the bay at dusk in a rising gale to attack the Marquis de Conflans' fleet sheltering there. Even from a comfy motor yacht bristling with electronics, these approaches can look forbidding in driving rain with seas breaking over Chaussée du Béniguet. Picture the scene in a winter blow aboard a lumbering three-deck warship under sail, with unpredictable tides, no GPS, no lights or buoys even, and probably rather dodgy charts.

This daring coup led to the famous victory for Hawke in the Battle of Quiberon Bay, a crucial turning point of the Seven Years War because Conflans was actually preparing to invade Britain. Troop ships had been assembling off the Morbihan for some time while Conflans was trapped in Brest by Hawke's blockade. But then sustained westerly gales caused Hawke to withdraw well offshore and finally to run for shelter in Tor Bay. By the time the weather eased and Hawke returned to his station, Conflans had escaped from Brest and was heading for the Quiberon rendezvous. Hawke caught up with the tail of Conflans' fleet just as they were entering Quiberon Bay and in the subsequent action the French lost five ships and almost 3,000 men, while Hawke lost *Essex*, *Resolution* and about 400 men.

Hawke's decisive action comes to mind whenever I'm cruising between Belle Ile and the Morbihan, especially while trying to pick up the elusive marks for La Teignouse. It's interesting to remember that 17th and 18th-century naval battles were fought with relatively crude gunpowder and shot that dictated the forms of attack and made fighting in confined coastal waters especially perilous. When Conflans withdrew his fleet behind the offshore reefs into Quiberon Bay, the only way Hawke could get into action was to follow him in, because the guns and cannonades of the time had very limited accurate range. The 32-pounder long gun could achieve an extreme range of about a nautical mile with a full charge, but a hit or miss at that distance was largely a matter of chance. You had to get in close to inflict serious damage, so Hawke couldn't cruise up and down outside the Quiberon shoals and take pot-shots at Conflans' ships. He had to lead his fleet inshore through some pretty miserly navigable gaps.

The Battle of Quiberon Bay
Dominic Serres
(Courtesy of the National Maritime Museum, London)

Ster-Wenn anchorage

any of the anchorages and make for Le Palais if necessary.

The largest-scale Admiralty chart covering Belle Ile is No. 2822, which is useful for approaching all the island's anchorages, although you have to be careful about interpreting detail as you get close inshore. The best detailed chart for exploring all the crevices around Belle Ile is the French SHOM No. 7142.

Ster-Vraz and Ster-Wenn The ruggedly spectacular fjord known as Ster Vraz faces west between steep cliffs at the northwest tip of Belle Ile, not quite a mile south of Pointe des Poulains. The entrance gap lies just over a mile southeast of Basse Occidentale des Poulains W Cardinal whistle buoy, but if arriving from this direction you should hold about half a mile offshore until the mouth of Ster Vraz is well open. Then make the final approach from more or less due west, keeping well to the south side of Ster Vraz on the way in, to avoid the various rocks off the northern point and off the north shore. The vertical cliff that forms the south headland of the entrance is easy to identify.

Ster-Wenn (sometimes spelt Ster-Ouen) is a narrow inlet in the cliffs that opens unexpectedly off the south side of Ster Vraz, a quarter of a mile in from the southern entrance point. It offers a dramatic anchorage in wild, timeless surroundings, but can actually become overcrowded during summer weekends. There is barely room to swing and the local method of mooring is to fetch up in the centre of the inlet and take stern lines ashore using the dinghy. A number of rings are set into the rocks for this purpose. Although Ster-Wenn seems

BELLE ILE TO GULF OF MORBIHAN

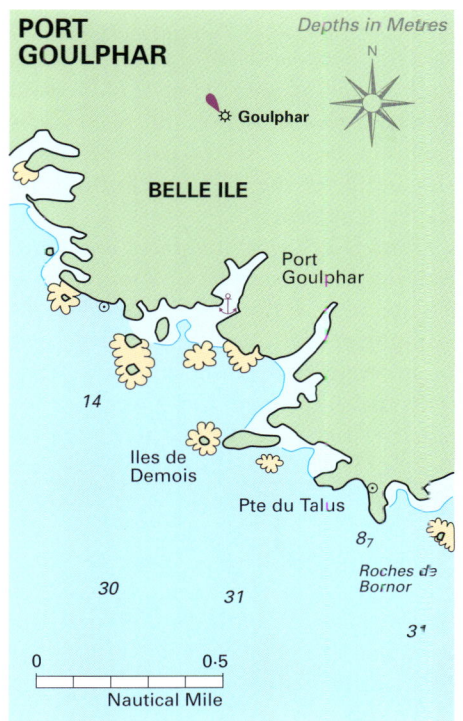

very sheltered once you are inside, strong onshore winds send in a dangerous surge and you must always keep a close eye on the weather and the underlying swell. It's not safe to enter or leave at night.

There is an attractive anchorage off the beach at the head of Ster Vraz, but you have to pick your way carefully past various rocks to reach it. Only continue beyond Ster Wenn near HW and so long as there is no swell, keeping about three-quarters of the way over to the south side. Edge in slowly, with a good lookout posted forward, and leave the obvious above-water rocks to port. The various drying and submerged rocks off the cliffs to starboard are usually easy to spot in the clear water.

Port Goulphar This is another craggy inlet in the cliffs, on the south coast of Belle Ile due south of Goulphar lighthouse. Although open to the

Opposite: Ster-Wenn anchorage

Belle Ile, west coast

SECRET ANCHORAGES OF BRITTANY 237

southwest and susceptible to any Atlantic swell, Goulphar is a splendid anchorage in quiet summer weather, especially in winds from between northwest through north to northeast. Stay a good half mile offshore until the entrance opens up and then approach from the SSW, with Goulphar lighthouse bearing between 010° and 015° for as long as you can see it. Hold this line until you have passed between the outer rocks and then bear to starboard, keeping to the middle of the inlet. Anchor outside the local moorings.

Port Kérel This inlet faces due south and lies three-quarters of a mile east of Pointe du Talut. The approach is straightforward, heading north true from a position about a quarter mile west of La Truie beacon tower. Anchor off the beach in the northeast arm of Kérel. The sandy bottom gives fair holding, but clear out if the wind comes onshore.

Port Herlin This is more of a bay than an inlet, just over a mile east of Port Kérel and open to the south. Port Herlin is a pleasant spot in quiet weather or winds from between northwest through north to northeast, so long as there's no Atlantic swell rolling in. Anchor about two cables off the beach in the middle of the bay, to avoid the various drying rocks fringing the shore.

Port de Pouldon The distinctive headland of Pointe de Pouldon juts into the Atlantic about one and a half miles

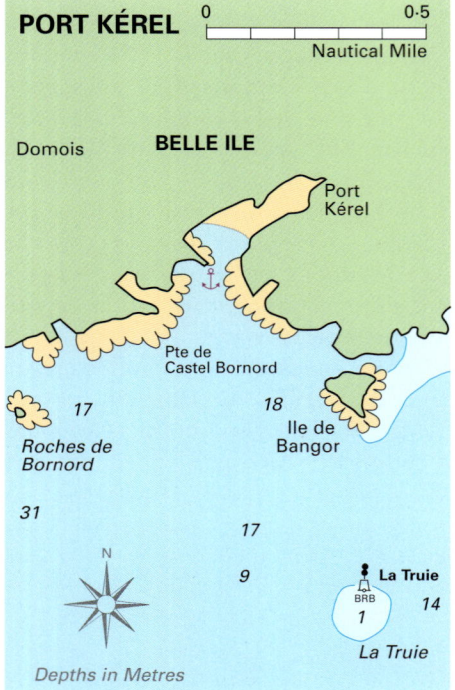

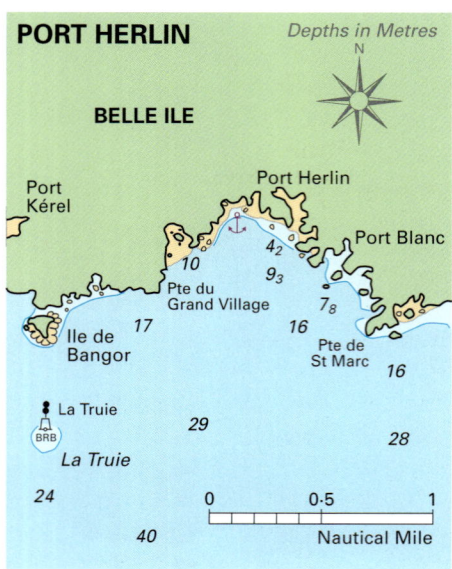

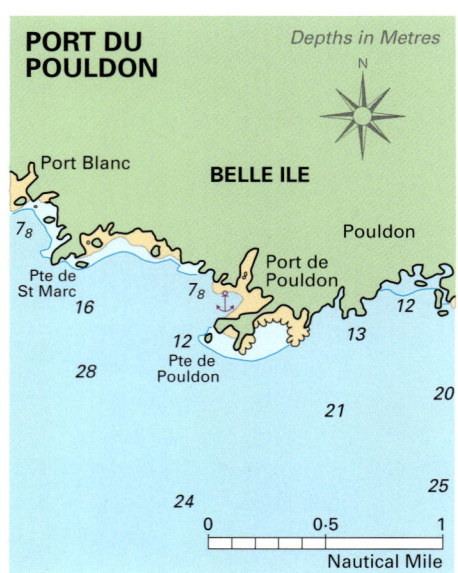

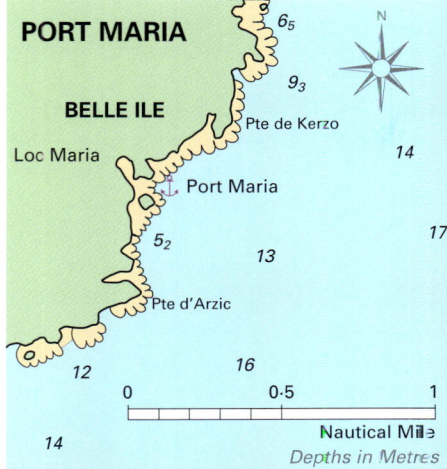
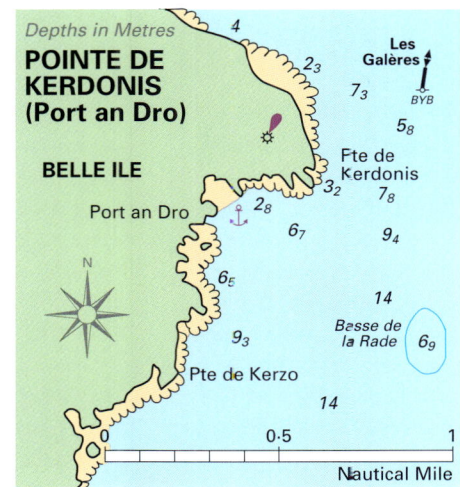

ESE along the coast from Port Herlin. On a quiet summer day, provided there is no onshore swell, you can anchor close north of this promontory and obtain good shelter from the northeast. Approach from the WSW, leaving Pointe du Pouldon and its off-lying rocks close to starboard. Make sure you avoid the drying rocks that extend seawards from the north side of the cove.

Port Maria An attractive anchorage off the southeast coast, sheltered from between west and northwest. Port Maria can be useful if you have arrived off Belle Ile from the direction of Ile d'Yeu and don't feel up to facing the rigours of Le Palais. Approach from the ESE, between Pointe de Kerzo and Pointe d'Arzic, allowing for the strongish tides off this corner of the island. Fetch up just outside the local moorings, over a sandy bottom. Port Maria has the advantage of being straightforward to leave at night if the wind should change, but it cannot be approached safely at night.

Pointe de Kerdonis (Port an Dro) This is the easternmost tip of Belle Ile and there is an attractive anchorage close south of the headland, well sheltered in westerlies or northwesterlies. Approach from the southeast, allowing for the strong cross-tides round this corner, and tuck well in towards the small beach known as Port an Dro. The cove has good holding over sand and the anchorage is easy to leave at night.

Port Yorc'h This is a useful anchorage on the northeast side of Belle Ile, only one and three quarter miles southeast from Le Palais and sheltered from between west through south to southeast. Entry is straightforward from the northeast, between Le Gros Rocher and La Truie N Cardinal beacon tower. Fetch up clear of the local moorings on the west side of the bay, i.e. closer to Le Gros Rocher than to La Truie. It's easy to leave Port Yorc'h at night and make for Le Palais if the wind should shift.

Port Salio and Port Guen Just west of Port Yorc'h, this wide curving bay between Pointe de Ramonet and Pointe du Gros Rocher is easy to approach and is pretty well protected from between west and south. However, it tends to be more susceptible than Port Yorc'h to swell and sometimes to wash from the mainland ferries. You need to tuck reasonably close in for the quietest water, but don't overdo it because there are some drying rocks close off the Port Salio beach. An alternative anchorage in the bay is Port Guen, a small inlet half a mile northwest of Salio. Le Palais harbour lies close to the northwest and is easy to reach at night if the weather

CITADEL AT LE PALAIS

Largest of the Brittany islands, Belle Ile takes the full brunt of Atlantic weather to provide a natural breakwater for Quiberon Bay and the rocky approaches to the Morbihan. Its main harbour, Le Palais, lies halfway along the landward side, once packed with sardine boats in the days of sail but now used mostly by yachts and the fast-moving ferries that roar in and out from Quiberon. On the north side of Le Palais harbour is an imposing star-shaped citadel, built for Henri II in the mid-16th century and later the stronghold of Nicolas Fouquet, who became Louis XIV's Surintendant des Finances in 1659. Fouquet bought the island from the Gondi family as a secure private refuge in case of 'misfortunes'. He became extremely wealthy and was able to maintain a sizeable private fleet to defend Belle Ile, but his increasing power and certain financial *confusions* rather put him out of favour with Louis, who eventually had Fouquet arrested and thrown into prison. (Those were the days.) In the late 17th century, after Louis had taken over Belle Ile, Le Palais citadel was heavily fortified by the ubiquitous Sébastien le Prestre de Vauban, who for 53 action-packed years was Louis XIV's chief military engineer. You can't cruise far around France without coming across one of Vauban's countless towers, harbour walls or forts. Vauban never did things by halves, and the citadel at Le Palais has immensely thick walls, double ramparts and corner bastions of prodigious strength. The powder magazine on the southwest side of the citadel was converted by Vauban from a virtually impregnable triple-walled tower that formed the hub of the original castle.

You can learn all about Belle Ile's turbulent history by visiting the citadel museum, an ideal retreat for a wet afternoon in Le Palais. Wandering those grandiose ramparts above the harbour, you can easily imagine the topsails of fully rigged ships of war appearing out of the mist and the thunder of heavy cannons from the citadel batteries.

The harbour and citadel at Le Palais

BELLE ILE TO GULF OF MORBIHAN

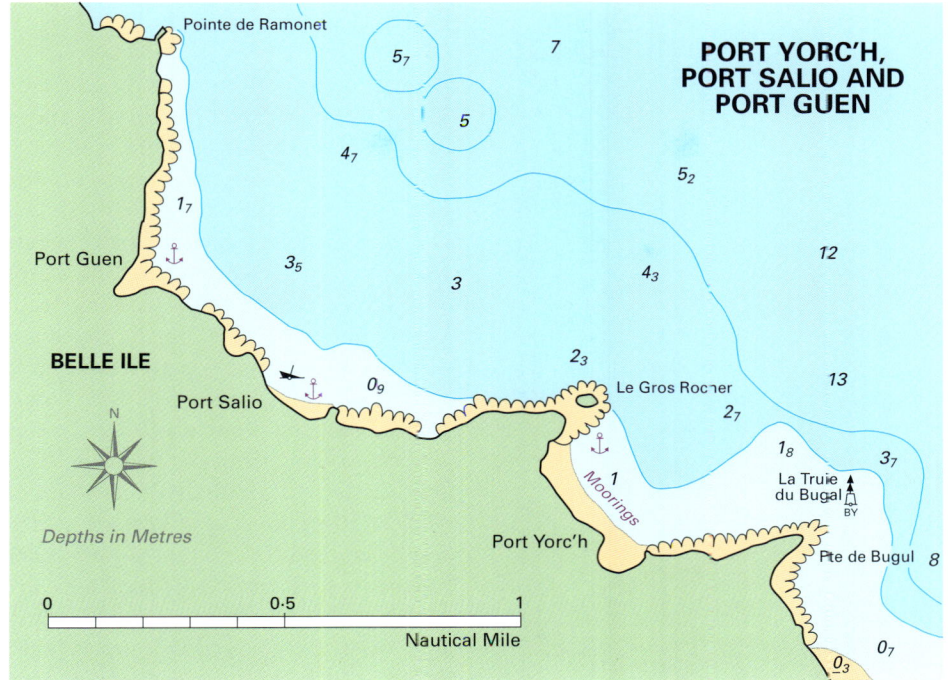

should change or a *vent solaire* springs up from the northeast.

Port Jean A small inlet on the north coast of Belle Ile, midway between Sauzon and Pointe de Taillefer. Port Jean is only suitable in quiet weather or if the wind is from a southerly quarter, but is worth bearing in mind if Sauzon is crowded. Approach is straightforward, finally coming in from the NNE and

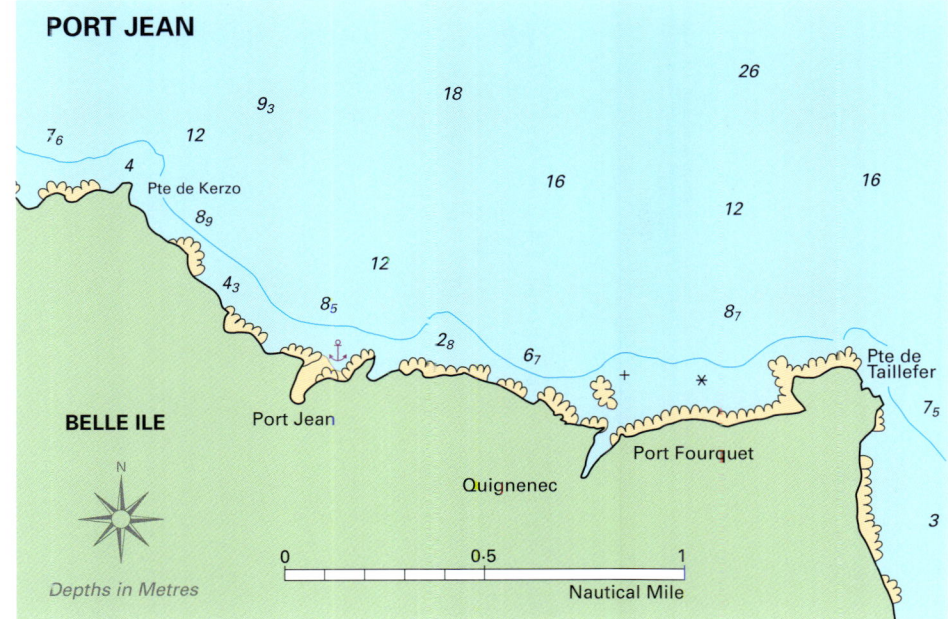

SECRET ANCHORAGES OF BRITTANY ⚓ **241**

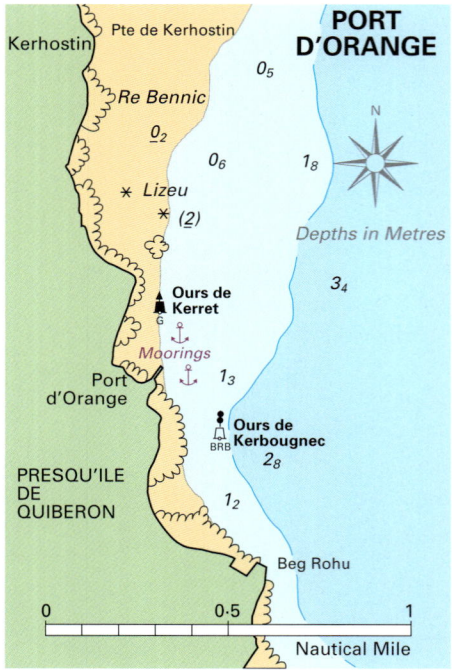

PORT D'ORANGE

A small drying harbour on the east side of the Quiberon Peninsula, two and a quarter miles NNW of Port Haliguen. There is a pleasant anchorage off the pier, open to the east and the *vent solaire* but with excellent shelter in any winds from the west. St Pierre village has shops and several restaurants. Coming from Port Haliguen, you need to stand offshore at least three-quarters of a mile to avoid the various rocky shoals between Basse Olibarte and Les Pierres Noires. Coming from the direction of La Trinité, steer to pass half a mile south of Men er Roué BRB spar beacon, which stands not quite one and a half miles northeast of Port d'Orange. There are oyster beds around this part of Quiberon Bay, marked by yellow and orange buoys, which should always be given a wide berth.

anchoring off a small beach. The cove is easy to leave at night if the wind should shift, when you can make either for Sauzon or Le Palais.

POINTE ST COLOMBAN

This low headland in the northwest corner of Quiberon Bay forms the east arm of the shallow approaches to the

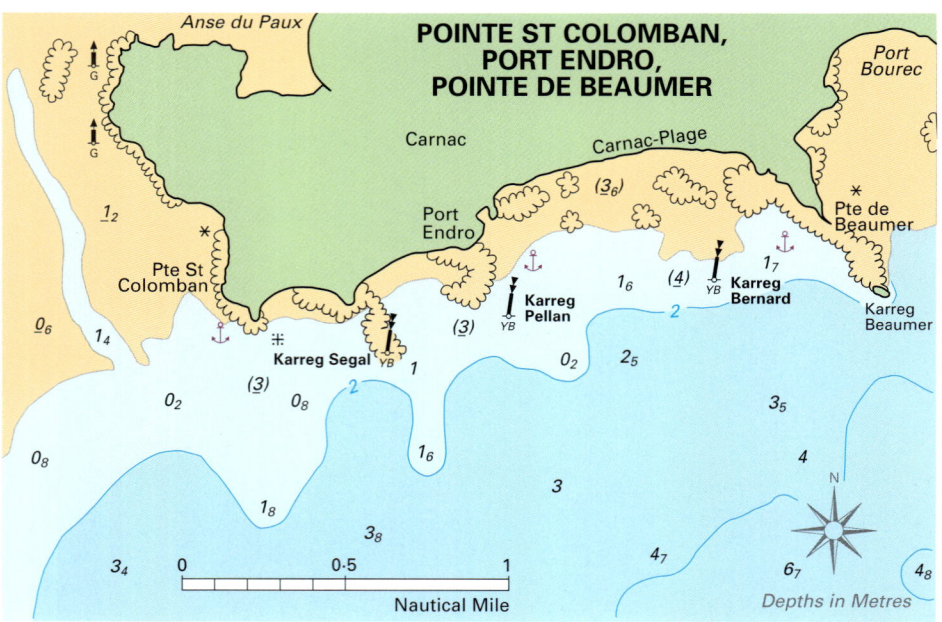

Anse du Paux (sometimes spelt Pô). At neaps, either in quiet weather or moderate winds with any north to them, there is a shallow anchorage close west of Pointe St Colomban, with good holding in muddy sand. Refer to Admiralty chart No. 2823 or French SHOM chart No. 7141.

CARNAC PLAGE (PORT ENDRO)

In quiet weather or northerlies you can anchor off Carnac Plage, fetching up a couple of cables northeast of Karreg Pellan S Cardinal spar beacon (sometimes spelt Carrec Pellan) and clear of the local moorings. Approach from due south, passing closer to Karreg Bernard S Cardinal beacon than to Karreg Pellan and then turning to port towards Port Endro. Sound carefully as you anchor, because the depths shoal quickly inside the two beacons. Refer to the French SHOM chart No. 7141.

POINTE DE BEAUMER

This low promontory lies a mile east of Port Endro and its line is continued southeastward by a narrow drying reef three and a half cables long. The reef terminates in Karreg Beaumer, an above-water rock 2.1m high, which forms the east arm of a shallow bay entered by leaving Karreg Beaumer to starboard and Karreg Bernard S Cardinal beacon to port. There is an anchorage here in offshore winds, just outside the local moorings with good holding over muddy sand. Refer to the French SHOM chart No. 7141.

RIVIÈRE DE ST PHILIBERT

This shallow unspoilt river flows into Quiberon Bay a mile east of La Trinité entrance and is interesting to visit around neaps in quiet weather. It's best to enter above half-flood. Proceed as though making for La Trinité, but turn off to starboard when Ar Gazek S Cardinal spar beacon bears 045° about half to three-quarters of a mile off. Steer northeast to leave this beacon a cable to port and continue past it for two cables on this heading. Then alter to the north towards Le Grand Pellignon red spar beacon, which stands on a rock right in the mouth of the river.

Enter the estuary by leaving Le Grand Pellignon beacon close to port and then head north by west to follow the channel, which is marked by withies. Anchor four cables upstream from Le Grand Pellignon, just east of Pointe de Bellec and clear of the moorings. This spot is well sheltered except in strong southerlies and has about half a metre at datum. The river is unlit and not safe to enter or leave at night. Refer to French SHOM chart No. 7141.

THE GULF OF MORBIHAN

Top: Morbihan anchorage south of Ile d'Arz

Bottom: The entrance to the Gulf of Morbihan

As you approach the Morbihan entrance, there's always a strong feeling of anticipation, of something interesting just round the corner: the coastline outside is not particularly dramatic. The east side has a long sandy beach curving round towards a picture-book promontory, Le Petit Mont, with an old dolmen at the top of its conical hill. Petit Mont forms the south arm of the entrance to Port du Crouesty, the largest marina on this stretch of coast and a possible base for leaving a boat between cruise legs. From Crouesty, the low grassy cliffs trend northwest towards Pointe de Port Navalo and its white lighthouse, the east arm of the Morbihan entrance.

On the west side, the low Locmariaquer shore is fringed with drying rocks and beacon towers. Approaching the entrance, you leave to port a distinctive hump-backed islet called Méaban, a useful landmark in hazy visibility. Further west, the coastline is indented by the shallow Anse de St Philibert (one of my favourite anchorages) and then the shallow buoyed entrance to the Crac'h River and La Trinité-sur-Mer. All this looks

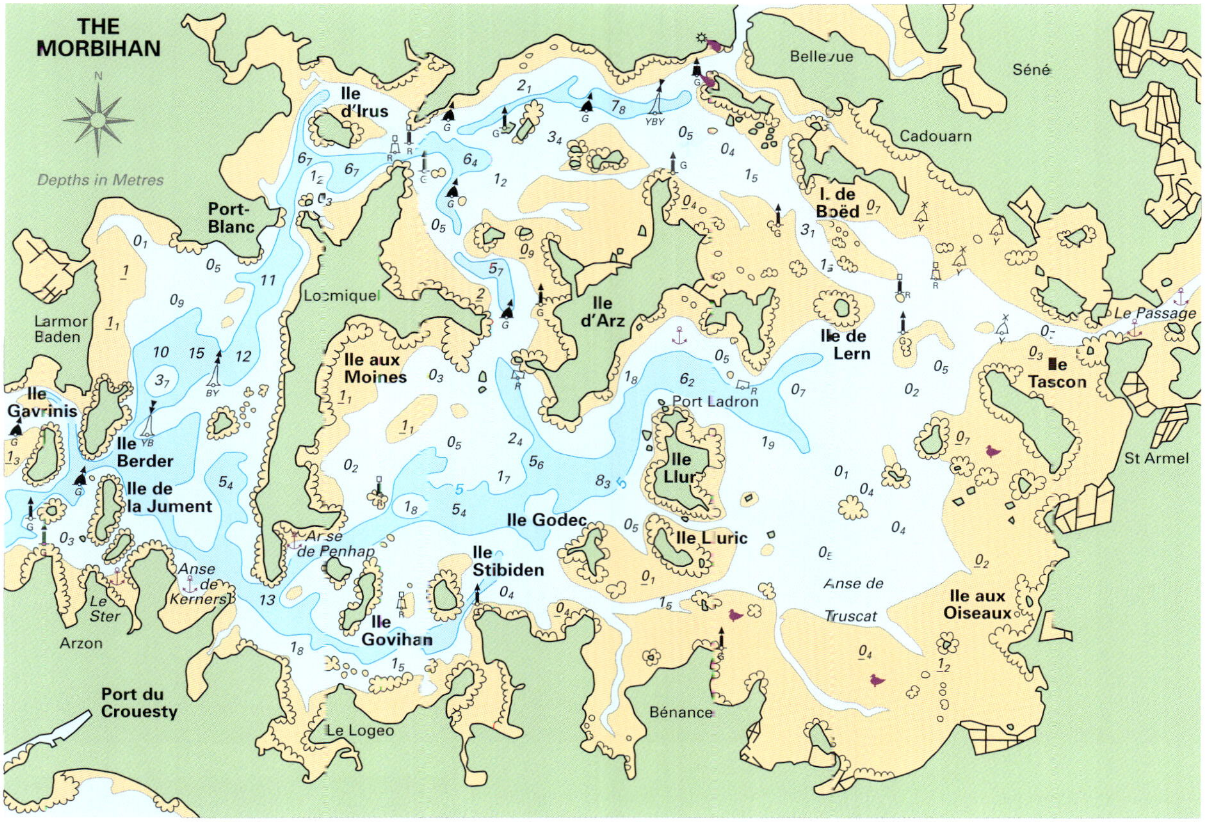

BELLE ILE TO GULF OF MORBIHAN

[Chart of The Morbihan — Depths in Metres]

pleasantly Breton as you arrive, but the interest level rises when you start to see inside the entrance.

You can sense intriguing waters by the sails appearing and disappearing behind islands. As you reach the tidal escalator and swirl in past Port Navalo, flat-bottomed shellfish boats cross your path, piled with mysterious baskets of oysters or mussels at critical stages. You must leave Grand Mouton green beacon well to starboard, allowing for the powerful flood that sweeps you northeast towards it. Preoccupied with Grand Mouton, and picking up Le Grégan tower and Petit Veisit white pyramid ahead, you are gradually aware of a kaleidoscope of islands opening up. The choice of channels is as tantalising as it is bewildering, and there's usually a flurry of hectic pilotage until you find your gap and go for it.

There are so many delightful anchorages in the Gulf of Morbihan that it's impossible to attempt to describe them all here. Neither is there any need, because the various pilot books cover a generous selection and you can easily find others with the help of the large-scale Admiralty chart No. 2371 or the French SHOM chart No. 7137.

Conleau narrows in the Vannes approaches

SECRET ANCHORAGES OF BRITTANY   245

A classic Breton lugger in the Auray River

Ile d'Arz ferry landing

One of the important considerations when choosing an anchorage in the Morbihan is to get out of the strong streams which run between many of the islands. Try to look for pronounced bays and then tuck as close inshore as you can; in most cases the bottom is muddy and there is no problem about cutting things fine and nudging aground for an hour or so around LW. At dead neaps you can reckon on a couple of metres above soundings at low water. If you choose a spot near local moorings, it's advisable to buoy your anchor. Always anchor well clear of the oyster and mussel beds marked on the charts.

The following anchorages are some of my favourites in the Morbihan.

Auray River This charming and inimitable 'oyster river' is well covered by the pilot books, probably the best-known anchorages being those in the

THE AURAY RIVER

I was first lured into the Morbihan by what has become a classic cruising book, George Millar's *Oyster River*. First published in 1963 and now reprinted, this nostalgic travellers' tale describes a leisurely summer cruise around the islands of the gulf and the glorious Auray River. As you sail up to Auray today – and you should try to sail as far as you can without reaching for the starter button – the very Breton atmosphere Millar described is still very much there. Of course there are rather more moorings now than during the 1960s, but you can still recognise the essential character of this timeless waterway:

'The Auray River, up which we had sailed five sea miles that first morning, bends through an arboreal landscape that is in fair weather smiling and in foul weather romantic with the flavour of R L Stevenson or Sir Walter Scott. . . . man has helped nature to make and preserve the scene. The banks are unvandalized. Nearly the whole length of either bank is privately owned by deep-rooted families. Châteaux, sizeable and old, are set back above the river with cattle grazing the parklands.'

The lower reaches of the Auray River are quite wide, although the west side is shallow and peppered with oyster beds. The channel follows the east side, and the sense of space is heightened by various fascinating gaps between islands. Opposite Ile Renaud and Sept Iles, the fairway narrows as it jinks round the islet of Grand Harnic (left to port as you go up) with the tiny entrance to the peaceful Anse de Badéne just opposite.

Above Grand Harnic, the Auray River is a mile wide until you reach Kerlevarec, where the channel narrows. It's easy to follow your way up using Admiralty chart No. 2371 or the French SHOM chart No. 7034, keeping between the moored boats. Two miles above Kerlevarec, the wooded banks edge close together before you reach the almost enclosed bight known as Le Rocher. If you can find an empty buoy up here, you'll be snug from any weather – an ideal retreat if gales are forecast. This is the highest part of the river that you can reach at any tide. Above Le Rocher is a wide shallow expanse where the channel divides. The east channel leads into a narrow gorge under a high (22m) road bridge to the village and drying harbour at Le Bono. You can sometimes moor on a trot just above the bridge, but the current is uneasy and the traffic noisy along the D101. Better just to nose up to see the picturesque village and its old iron bridge, before turning back and taking the north channel towards Auray.

This last stretch is only navigable above half-tide, marked mainly by starboard-hand spars. Just below Auray you reach another busy bridge (14m headroom), with visitors' moorings just before it. Between springs and neaps, though, I usually press on under the bridge to the harbour narrows below St Goustan. This charming old village is on the east shore and many of its timbered houses date back to the 15th century. There are pleasant trot moorings off the west bank, opposite Le Relais Franklin. Don't stay alongside the quays though, since these dry to sticky mud, and don't moor too far up near the old stone bridge, where the currents are stronger.

In the days of sail, St Goustan was one of the busiest ports in Brittany, with trading coasters and tunny boats drying out alongside.

There are restaurants and small shops near the quayside, but St Goustan is something of a tourist trap and you'll do better eating up in Auray. Leave the dinghy at the pontoon on the east shore, walk along the east quay, across the stone bridge, turn downstream a little and then follow a winding path up to a vantage point above the harbour. From here you can cut inland (west) to the centre of Auray, a splendid old market town with good shops, a very French market square, the fine 17th-century church of St Gilda and bistros aplenty.

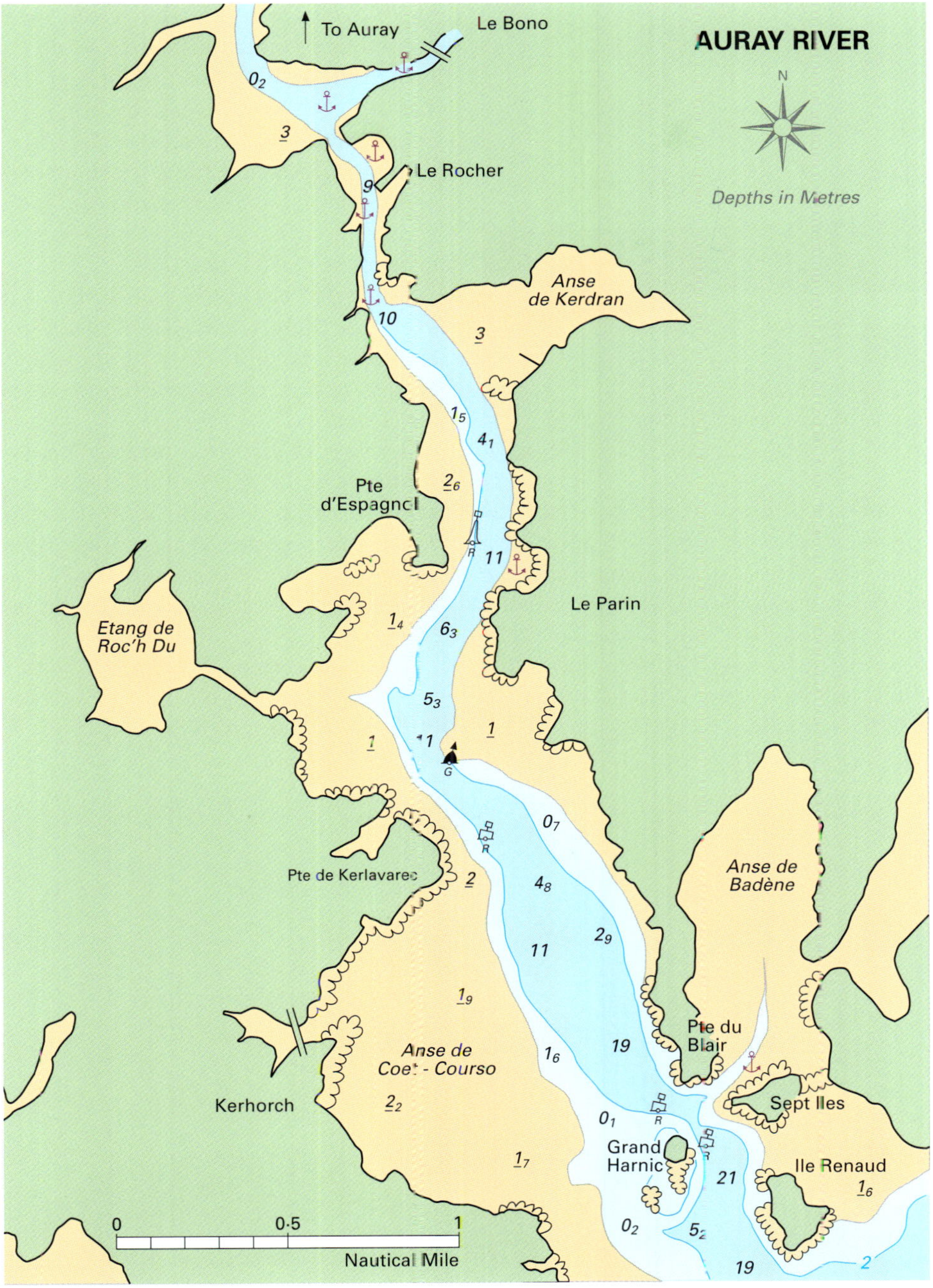

BELLE ILE TO GULF OF MORBIHAN

attractive wooded narrows opposite Le Rocher. You can also fetch up in Le Bono creek, on the starboard hand just above Le Rocher, although the holding here is unreliable over the rocky bottom. Further downstream, there is an anchorage on the east side of the river, in the small bay opposite Pointe d'Espagnol; you can tuck in close here at neaps, nicely out of the stream. Time passes slowly, marked only by the ebb and flow of the tide.

SECRET ANCHORAGES OF BRITTANY 249

Anchored off Ile d'Arz in the Gulf of Morbihan

Approaching Conleau narrows in the Morbihan

Anse de Badène At neaps, boats with a modest draught can edge a little way into the Anse de Badène, turning northeast out of the lower Auray estuary opposite Grand Harnic island. The tide is strong in the gut between Pointe du Blair and Sept Iles and it's best to enter near low water, watching the echosounder carefully as you go. Keep well north towards the outer green spar beacon and Pointe du Blair to avoid the rocky shoals off the west tip of Sept Iles. The Anse de Blair dries, except for a narrow tongue extending northeast from the mouth. Anchor about a cable north of Sept Iles, or sometimes you can find a vacant mooring. No facilities here – a taste of the real Morbihan.

Anse de Penhap At the southern tip of Ile aux Moines, this shallow bay is my favourite landing for visiting the island. Neaps are best, tucking well in to escape the current.

Le Ster Near the entrance on the south shore of the Gulf, Le Ster is a shallow bay just south of Ile Denton. This spot is sheltered from everything except fresh northwesterlies. Anchor or find a buoy a cable east of Pointe de Bernon.

Anse de Kerners A larger bay half a mile east of Le Ster, on the opposite side of Pointe de Kerners. Near neaps you can tuck well in here.

ISLANDS OF THE MORBIHAN

There are something like 50 islands dotted around the Gulf, all private except the two largest – Ile aux Moines and Ile d'Arz. Most of the smaller islands are uninhabited, but some have secluded holiday homes tucked behind sheltering pines. An ancient custom permits strangers to land on the shores of private islands in the Morbihan, so long as they stay close to the high-water mark. Even so, you should exercise discretion about intruding into these mouth-watering private domains.

Ile Berder, near the entrance to the Gulf, has a large convalescent home, and it's usual to ask permission before exploring the island. In westerlies, we've often anchored off a shallow bay on the northeast side of Berder.

Ile aux Moines and Ile d'Arz are both worth visiting, although Arz is my favourite. Moines is the largest island in the Morbihan (three miles from north to south) but the most visited by tourists, who pile across in the frequent ferries from Port Blanc. Ile aux Moines has a small marina on its northwest side, but the berths are shallow at low water and disturbed by wash from the ferries. To sample the flavour of Moines, find an empty corner in the Anse de Penhap, right at the south end of the island. There's still room to anchor (tuck well into the shallows out of the current) or you may find an empty buoy. Landing at the small jetty there, you can wander north through Penhap village and into the quieter corners of Moines that tourists rarely penetrate.

Ile d'Arz is a delightful retreat, low and sandy with fascinating unexpected vistas across the Gulf in all directions. It's best to moor off the wide bay that forms the south coast of Arz, where the jetty is only 10 minutes' stroll from the village, Le Bourg. This bay is shallow a long way out, so neaps are best for staying a day or two. There's anchoring room outside the moorings, or you may find an empty buoy. The Glénans sailing school has a base on the south shore of Arz, so you'll often see dinghies and catamarans whizzing about with great flair.

As you wander up towards Le Bourg along sandy paths, the outside world may never have existed. Surrounded by tidal water, vast mudflats and acres of oyster beds, Arz seems an idyllic time capsule cut off from metropolitan concerns. The ornate spire of the church is a focal point as you stroll. The tiny post office is cool for writing postcards and the café in the main street is a convivial gathering place for quenching your thirst. You can genuinely unwind on this sleepy island. Anyone who takes a mobile phone anywhere near Arz needs certifying.

Island ferry landing

The church on Ile d'Arz

Le Passage

night and make for Crouesty, so long as you head SSW at first to make a safe offing from Le Bec du Colombier and Petit Mont. Refer to Admiralty chart No. 2371.

ANSE DE CORNAULT

Just over four miles southeast of the entrance to the Morbihan, Pointe du Grand Mont is a prominent headland of steep cliffs with a disused semaphore building on its west edge. The Anse de Cornault is a shallow sandy bay about three-quarters of a mile north of Pointe du Grand Mont, providing a pleasant anchorage in settled weather with the wind from between east and northeast. As with La Grève de Feugeot, any onshore swell tends to make itself felt in the shallow water over a gradually shelving bottom.

Although there are various shoals in the offing, the approach to Anse de Cornault is not difficult from either direction along the coast. Coming from the Morbihan, make for Pointe du Grand Mont until Grand Rohu islet, just off the coast, is just abaft the port beam. Then head ENE for the centre of the bay. Coming from the east, say from Penerf or the Vilaine River, it's best to keep south of St Jacques and Basse du Grand Mont S Cardinal buoys before turning north for the Anse de Cornault.

Coming from seaward, make for Pointe du Grand Mont, which is easy to pick out. However, don't confuse La Chimère and Basse du Grand Mont S Cardinal buoys – the former can be passed either side in quiet weather, but the latter must be left to the east to avoid the rock, almost awash at datum, that it guards. Basse de St Gildas W Cardinal buoy is a useful mark, one and a half miles southwest of Anse de Cornault on the seaward side of a small rocky shoal with 0.9m over it at datum.

Port Ladron Three-quarters of a mile southeast of the south landing jetty on Ile d'Arz, Port Ladron is a shallow sandy bight between the north end of Ile d'Ilur and the off-lying islet of Le Charles. Idyllic for pottering with the family.

Le Passage At the far east end of the Morbihan, the peaceful sound north of Ile du Passage is well off the beaten track. Follow the shallow channel eastwards from Arz above half-flood. There are buoys and room to anchor off Le Passage. From the landing on Ile du Passage, you can stroll south to St Armel village, where there are a few shops and cafés.

LA GRÈVE DE FEUGEOT

A little way outside the Morbihan, a mile east of the entrance buoys for Port du Crouesty marina, is the wide sandy bay known as La Grève de Feugeot. The west headland of this bay, Petit Mont, has a rocky spur on its east side, known as Le Bec du Colombier, but there's an anchorage off the beach three cables or so east of these rocks, sheltered from between north and east. The holding is good in sand and mud, but any onshore swell tends to be amplified because the bay is shallow and gradually shelving. It's feasible to leave this anchorage at

You can leave Anse de Cornault at night with care and Port du Crouesty can easily be reached by standing out due west until the white sector of Port Navalo light appears. Then you can turn due north towards Crouesty or the Morbihan entrance.

USEFUL ADMIRALTY CHARTS

No. 2371 Golfe du Morbihan
No. 2822 Ile de Groix to Belle Ile
No. 2823 Quiberon to Croisic

USEFUL SHOM CHARTS

No. 7137 Golfe du Morbihan
No. 7141 Baie de Quiberon
No. 7142 Belle Ile

Traditional Breton cottages up at Le Passage

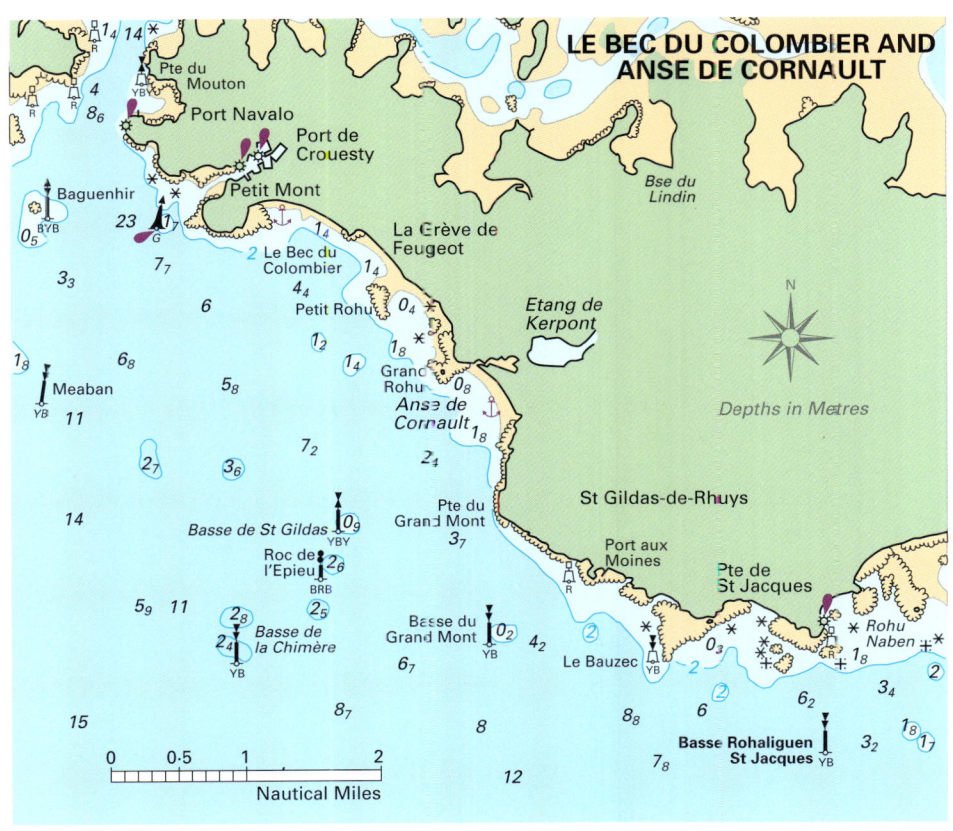

The anchorage at Tréac'h er Goured on Houat

SECRET ANCHORAGES OF BRITTANY   255

CHAPTER 8

HOUAT AND HOËDIC TO THE VILAINE ESTUARY

The wild twins of Houat and Hoëdic, which help protect Quiberon Bay from Biscay swell, seem to take you back in time even more than small islands usually do. Houat is the more westerly of the two, shaped on the chart like a lobster, with a highly individual population of just over 400. Hoëdic is smaller, but no less independent. Both rely on a simple economy of fishing and the tourists who come over from Quiberon, Port Navalo and Vannes for the peaceful atmosphere and the mouth-watering beaches.

Houat's colourful harbour, Port St Gildas, lies on its north coast and is home to about 40 local fishing boats that line the quays. On summer weekends, yachts pile in around them in a sociable mêlée. A small road winds up towards the village centre, with its low, whitewashed houses and solid, 18th century church. The magnificent long beach of Treac'h-er-Goured has its back to the prevailing winds, a paradise of silver sand backed by rolling dunes covered with marram grass and tamarisk. A tiny drying harbour is tucked into the south corner of this spectacular bay. Houat has a tempting selection of anchorages to suit most winds and I can think of no finer tonic than a few days spent lying off its charming coast, venturing ashore from time to time for fresh bread and a spot of exercise.

Hoëdic is half the size of Houat and has few inhabitants, fewer ferries and fewer visitors, but still has its own *mairie*, a village school, post office and the splendid 19th century church of St

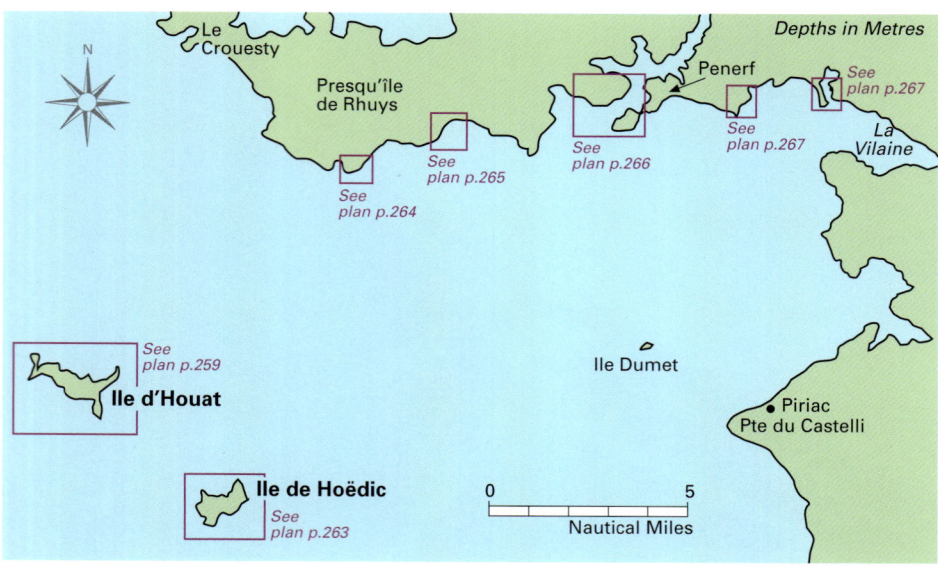

256 SECRET ANCHORAGES OF BRITTANY

Houat, Tréac'h er Goured

Goustan. Hoëdic has some idyllic anchorages too, but being generously surrounded by rocks and shoals, they are slightly more delicate to enter or leave. Its main harbour is on the north coast, but there is also a small drying harbour, Port de la Croix, in its southwest corner.

Back on the Brittany mainland, the coastline to the east of the Morbihan entrance has a few anchorage hideaways between St Jacques and the Vilaine River, with the upper reaches of the Penerf estuary as the most sheltered and attractive. This stretch of south-facing shore looks a shade hostile on the chart, with rocks and shoals extending well out, but when you explore here in quiet summer weather it is actually rather attractive, with some interesting changes of mood and scenery along its length.

These waters are partly sheltered by the natural breakwater provided by Houat, Hoëdic and the string of reefs stretching back to the Quiberon peninsula.

Penerf River

In this chapter I meander as far east along the mainland coast as Kervoyal and Billiers, which are both on the north side of the shallow mouth of the Vilaine. These two anchorages near the river entrance are interesting in themselves, but can also be useful if you arrive off the Vilaine fairly late at the end of a day's cruising and would rather get tucked up in good time for dinner than continue upriver that evening. Kervoyal is simple to approach at any state of tide. Billiers offers the best shelter, but should be entered on the last of the flood and is most suitable for bilge-keelers.

Delightful though this whole area is, you need a certain amount of care in hazy summer conditions, either cruising around Houat and Hoëdic or when making passages between the islands and the mainland. Although the tides are generally moderate, some strong local sets can be experienced in unexpected directions, which may catch an over-relaxed navigator unawares.

From the Morbihan entrance, you can reckon about 10 miles to the north side of Houat and not quite 13 miles to the north side of Hoëdic. Both these passages are simple in clear visibility and preferably with at least a couple of hours' rise of tide. But the final approach to these islands should be from more or less due north. At springs with calm winds, guard against being set to the west of either Houat or Hoëdic if, as will be likely if arriving from the Morbihan, you are approaching on the ebb. Both islands have rocky dangers on their northwest sides.

When sailing northeastwards from Houat or Hoëdic towards Penerf or the

HOUAT AND HOËDIC TO THE VILAINE ESTUARY

Vilaine, remember that, although Plateau de la Recherche has plenty of water over it most of the time, there are several isolated heads which would be close to your keel at low water. The most easterly has only 1.8m over it at datum and is two miles from the nearest buoy, the BRB isolated-danger Locmariaquer buoy which guards the south edge of this four-mile string of shoals. The west end is marked by the Recherche W Cardinal buoy.

The entrance to Penerf lies 13 miles northeast from Houat or Hoëdic and about five miles west of the mouth of the Vilaine. There are quiet anchorages in its upper reaches and the approach to this gracious river is much simpler than most pilot books suggest, so long as you have reasonable visibility and at least three hours rise of tide. The outer approaches to the Vilaine estuary are wide and shallow, with rocky dangers on the west side off Penerf. Other factors being equal, it's best to arrive at the entrance soon after half-flood when going in, and as near high water as possible when coming out. In fresh onshore winds, especially from the southwest, the mouth of the Vilaine should be avoided. A spring ebb in these conditions can kick up a devilish sea.

HOUAT

This small, curiously-shaped island lies 10 miles south of the entrance to the Morbihan and about seven miles ENE of Le Palais. Barely two and a half miles from end to end, Houat is part of that geological tangle of reefs and islands that straggles southeast from Presqu'île de Quiberon for nearly 15 miles as far as Les Grands Cardinaux. Although Houat has a snug harbour at the east end of its north coast, this is home to a considerable fleet of local fishing boats and usually seems full even before yachts start arriving. However, the shape of the island is such that you can almost

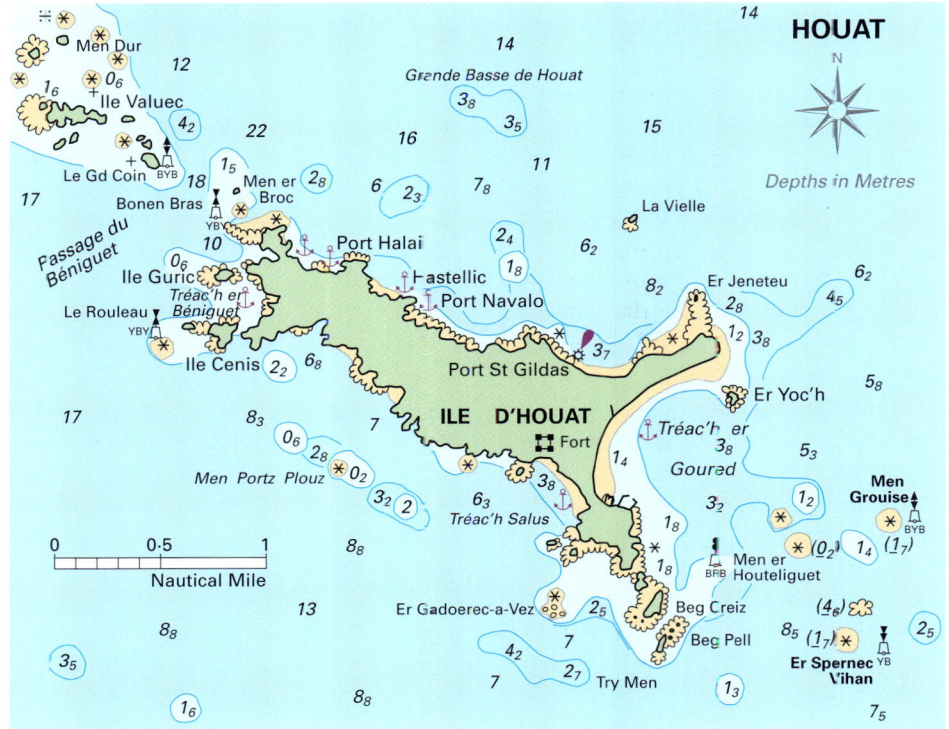

SECRET ANCHORAGES OF BRITTANY   259

Houat NW coast

always obtain shelter by anchoring off one of its splendid beaches.

Tréac'h er Béniguet This attractive bay at the west tip of Houat is a good spot in easterly weather. The approach is straightforward from the west, leaving Le Rouleau W Cardinal beacon tower to starboard and then steering for the middle of the bay to pass north of Ile Cenis and south of Ile Guric. Keep clear of the various drying rocks between Le Rouleau and Ile Cenis and also those that extend southwest from Ile Guric. Fetch up close off the beach in about two metres at datum.

If arriving from the north through the Passage du Béniguet, pass between Le Grand Coin E Cardinal and Bonen Bras W Cardinal beacon towers and then make good about 220° until Tréac'h er Béniguet is well open before turning east towards the bay. The anchorage is not safe to enter or leave at night because the west end of Houat is completely unlit. For any detailed Houat pilotage use either Admiralty chart No. 2835 or the equivalent French SHOM chart No. 7143.

Tréac'h Salus This bay on the south side of Houat is formed by the west shore of the narrow sandy peninsula that extends SSE from the island to form the southerly 'claw' of its lobster shape. Tréac'h Salus is sheltered from the northeast, so it's a good spot when the *vent solaire* is in evidence. The approach is not difficult but needs a little care, because a string of unmarked shoals lurks a quarter of a mile off the island's south coast. Most of these heads are covered (some only just) at datum, but the central rock – Men Portz Plouz – dries 1.5m. Ideally, you should approach Tréac'h Salus near high water, when the streams are slack and you have plenty of depth over these dangers.

Coming from Le Palais on Belle Ile within a couple of hours of high water, you can make good a direct line of 077° from Le Palais pierheads, heading for the highest point of the southeast extremity of Houat on this course. This track passes just over half a mile northwest of Le Pot de Fer BRB isolated danger buoy. Carry on towards the southeast promontory of Houat until you are half a mile off and then edge to port for the middle of Tréac'h Salus beach.

Coming from the northwest or from the north via the Passage du Béniguet, it's best to keep a mile off the south coast of Houat until you are safely past the Men Portz Plouz shoals. Then you can turn in towards Tréac'h Salus when the middle of the beach bears 060°.

Coming east-about the island from Tréac'h er Goured, preferably near high water, pass close either side of Men er Houteliguet isolated danger beacon tower and then skirt by a good quarter of a mile the tail of above-water rocks forming the southeast tip of the island. Once round the outer rock, Try Men, make good due west true until the old fort on Houat bears well east of north, to be sure of clearing the various rocks on your starboard hand before turning up for Tréac'h Salus.

Tréac'h Salus is exposed to the southwest and you should leave the anchorage in daylight if there's any risk of the weather picking up from that direction. Navigation anywhere in the vicinity of Houat is not safe at night. The holding in Tréac'h Salus is reasonable in moderate conditions, but the bottom is sand and rock rather than reliable mud. Admiralty chart No. 2353 is OK, but the French SHOM chart No. 7143 is best.

Tréac'h er Goured This wide sandy bay on the east coast is perhaps the most popular of the Houat anchorages and many French yachts sail out at weekends from Crouesty or the Morbihan. Tréac'h er Goured is sheltered from between northwest through west to southwest and the holding is fair if you tuck close in. The approach is easiest from the north, rounding the northeast promontory of the island and then Er Yoc'h islet (18m high). Coming from the east, say from the Vilaine River, leave Men Groise E Cardinal beacon a good three cables to the south before leaving Er Yoc'h to starboard and entering the bay. There are numerous dangers in the offing to the southeast of Tréac'h er Goured.

It's not safe to approach the anchorage at night, but you can leave with care by sailing out to the northeast between Er Yoc'h and Men Groise beacon (unlit). Once you are clear of the island, it's easy to make for Port du Crouesty, nine miles away to the north. The only light on Houat is the breakwater head at Port de St Gildas.

Port Navalo One of the smaller bays towards the west end of Houat's north coast, Port Navalo lies just east of the short headland known as Er Hastellic, a mile from Port de St Gildas. The approach is straightforward, either from the north, from the west via the Passage du Béniguet, or from St Gildas itself. Navalo is snug in southwesterlies, although at night during fine summer spells it is open to the *vent solaire*. The holding is fair over a sandy bottom. You can safely leave the anchorage at night and make for St Gildas or Crouesty.

Hastellic This small bay, with its attractive beach, lies just west of Port Navalo, between Grunn er Vilaine and Er Hastellic. The approach is similar to that for Navalo, except that you fetch up on the opposite (west) side of Er Hastellic.

Hoëdic, Port de l'Argol

Port Halai and Port Ler These two coves are next to each other at the northwest tip of Houat, close east of the headland called Beg er Vachif. They are partly separated by a drying rocky spur, so make sure you are definitely in one cove or the other. There is good shelter here in southerlies, although the *vent solaire* can be a nuisance at night during warm weather spells. You can clear out at night by heading straight offshore at first and then making for St Gildas or Crouesty, but take careful account of the tides off this corner of the island. In particular, beware of being set west towards the dangers fringing the Passage du Béniguet.

HOEDIC

The tiny island of Hoëdic lies just over three miles southeast of Houat, smaller and more regular in shape than the latter, and fringed with dangers except off its north side. There are two small harbours: Argol on the north coast where the ferries land, and Port de la Croix on the south coast. Argol is tiny and packed with local boats, while Port de la Croix dries right out and is almost as crowded. The best charts for Hoëdic are either Admiralty 2835 or the French SHOM chart 7143.

North coast The simplest north coast anchorage is off the beach to the west of Argol, near the old lifeboat slip. The approach is not difficult from due north, leaving Er Rouzes E Cardinal beacon tower a mile to the west while making for the west half of the island. You can also fetch up off Argol itself, provided you keep the east pierhead bearing east of south on the approach and when you anchor. Not quite half a mile north of this pier is La Chèvre, a patch of drying rocks marked by an isolated danger beacon. It's possible to leave either anchorage at night with care, using Argol pierhead light to keep you clear of La Chèvre and the dangers northwest of Hoëdic.

HOUAT AND HOËDIC TO THE VILAINE ESTUARY

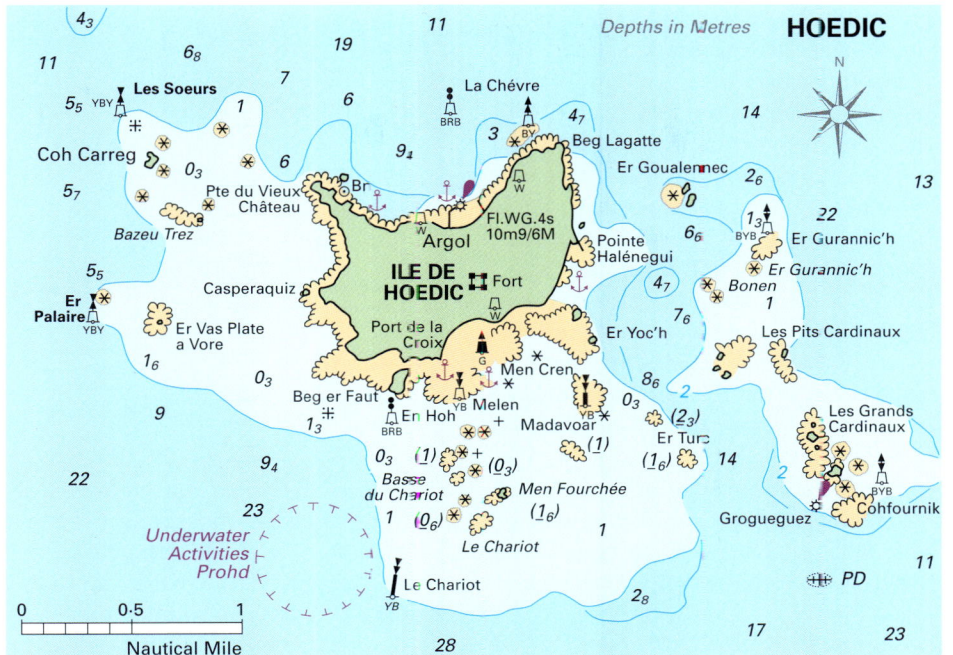
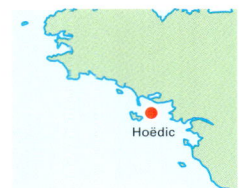

Pointe Halénegui There's a secluded anchorage on the southeast side of Hoëdic, just off the beach between Pointe Halénegui and Er Yoc'h bras, an above-water rock a quarter of a mile south of Pointe Halénegui. The easiest approach is from the north near high water. Leave the northeast tip of the island, Beg Lagatte, two cables to the west, keep a similar distance off the east coast and pass inside Er Goualennec islet.

If you do arrive near low water, it's important to avoid two rocky patches only just covered at datum, which lurk up to two cables east of Pointe Halénegui. Otherwise, round the headland this distance off and turn to starboard towards the anchorage, being sure to avoid the drying spur that extends SSE from Pointe Halénegui for about a cable. Do not stray south of Er Yoc'h bras, as this is only the above-water tip of a wide plateau of drying rocks stretching SSE for over a quarter of a mile.

An alternative approach is from the northeast, preferably near high water, passing between Er Goualennec islet and Er Gurannic'h E Cardinal beacon tower. Bonen shoal, with 1.3m over it at datum, lies a cable north by west of the beacon tower, but carries sufficient depth for most yachts in quiet conditions, except near low springs. Head southwest for Er Yoc'h bras and turn into the anchorage as the beach starts to bear north of west.

The Halénegui anchorage is an idyllic spot if the weather is right, sheltered from between northwest and west, and partly protected from swell by Er Yoc'h bras plateau and the various drying rocks south of Hoëdic. It's feasible to stay here overnight so long as the forecast looks OK. Leaving after dark is not advisable, although just about possible on a clear night by following the coast northwards two cables off. The only light is Les Grands Cardinaux, one and a half miles southeast of the anchorage, which can give you

SECRET ANCHORAGES OF BRITTANY 263

somewhat fine clearing lines for Er Goualennec islet and Beg Lagatte.

South coast There's a fair weather anchorage outside Port de la Croix, with good shelter in northerlies and from the *vent solaire* overnight. The approach looks tricky on a small-scale chart, but is actually fairly straightforward near high water provided conditions are calm. From a position between half and three-quarters of a mile due south of Les Grands Cardinaux lighthouse (red and white tower, 28m high) bring Madavoar S Cardinal beacon tower in transit with the right-hand edge of Hoëdic fort bearing 320° and follow this line shorewards. Do not confuse Madavoar with Roche Melen S Cardinal beacon tower, which stands half a mile further west.

As you draw near Madavoar, leave it to starboard and curve gradually to port towards a position about half a cable south of Men Cren green beacon tower. Fetch up here, take a careful sounding and work out your low water depth. At neaps you can stay afloat nearer the harbour entrance, passing between Men Cren and a red spar beacon. Before staying overnight, you need to be pretty sure that the wind won't shift anywhere into the south. It's not safe to leave this anchorage after dark but, in *extremis*, you could enter Port de la Croix above half-flood and try to find a vacant stretch of quay to dry out alongside.

ST JACQUES-EN-SARZEAU

Back on the Brittany mainland, St Jacques is a small drying fishing harbour at the southern tip of the Sarzeau peninsula, close east of Pointe de St Jacques and nine miles NNE of Hoëdic. There is a neap tide anchorage off the pierhead in quiet weather or in offshore winds from between west through north to northeast. It's easiest to approach above half-flood because, near low

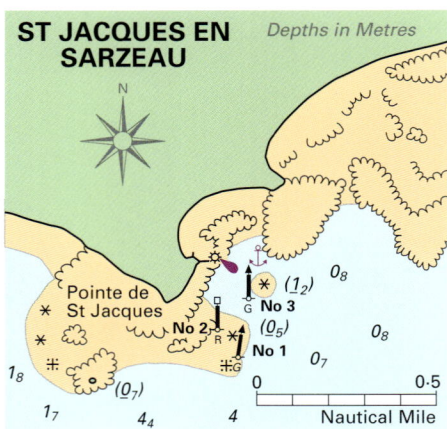

water, care is needed to avoid Basse de St Jacques – a rocky unmarked shoal a mile southeast of Pointe de St Jacques – and also the drying ledges that lie up to half a mile southwest and south of the point.

Coming from the west with plenty of rise of tide, pass a quarter of a mile south of Le Bauzec S Cardinal beacon tower and then make good due east true along the coast, aiming to leave St Jacques S Cardinal buoy a quarter of a mile to the south. Near low springs you would need to pass south of this buoy for safety, but at springs the St Jacques anchorage would be too shallow anyway and you wouldn't be here. Now head NNE keeping Pointe de St Jacques and its off-lying red beacon broad on the port bow. Turn for St Jacques harbour pierhead when it bears 310°.

Coming from the east you also need plenty of rise of tide, because near low water it can be tricky to steer a safe course between Basse de St Jacques and the various shoals that straggle well offshore to the east of St Jacques. But above half-flood you can come along the coast making due west true on a line that would leave St Jacques S Cardinal buoy a quarter of a mile to the south. Before you get there, turn inshore for St Jacques harbour pierhead when it bears 310°.

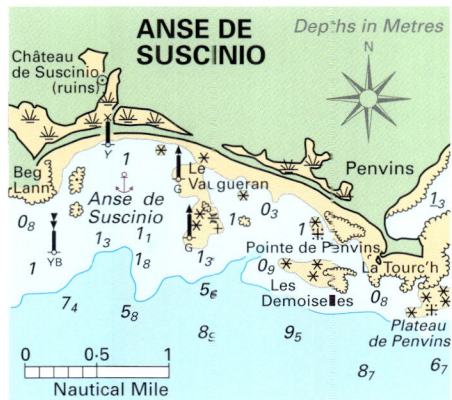

Fetch up about 50 metres southeast of the pierhead or edge a bit further in if the depth allows. But when choosing a spot, and indeed when coming in, be sure to stay clear of the drying rocks that come well out from the east side of the bay. A riding light is advisable if you are anchoring overnight, because of local fishing boats coming and going. You can enter or leave St Jacques at night with at least two hours rise of tide by keeping the pierhead light bearing 310° until you are safely clear offshore. The French SHOM chart No. 7135 is the best for this anchorage and the coast to its east.

ANSE DE SUSCINIO

This wide shallow bay lies three miles east along the coast from Pointe de St Jacques and makes a pleasant anchorage in quiet weather or offshore winds, so long as there is no onshore swell rolling in. There are various rocky shoals within and either side of the bay and it's preferable to enter within a couple of hours of high water. From about two miles off, bring the prominent Château de Suscinio to bear 345° and approach the coast slowly on this bearing, keeping a close watch on the echo-sounder as you come within a mile of the beach. As you come abreast the first of two green spar beacons (each left about three cables to starboard) the depths should shoal slightly as you cross a 1.3m patch. Then the depth should increase again as you draw into the bay.

The second green beacon stands at the head of the bay to starboard. Anchor just before this beacon bears due east true, or edge a little further inshore if your draught and the tide allow. The anchorage is especially snug in northwesterlies, but you should clear out in good time if the wind looks like coming onshore. It's not safe to enter or leave the Anse de Suscinio at night. The French SHOM chart No. 7135 is best for this anchorage.

Summer pottering in the Pénerf estuary

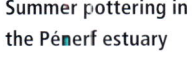

PÉNERF

This attractive unspoilt river is well covered by the pilot books, but it's worth pointing out that the shallow upper reaches opposite Pen-Cadenic usually offer better shelter and seclusion than the moorings or anchorage off Pénerf itself. Beyond Pen-Cadenic the river widens and becomes very shallow, although shoal draught boats can venture up the eastern arm near high water and find one or two stretches of foreshore where it's possible to take the ground.

Pénerf and Pen-Cadenic

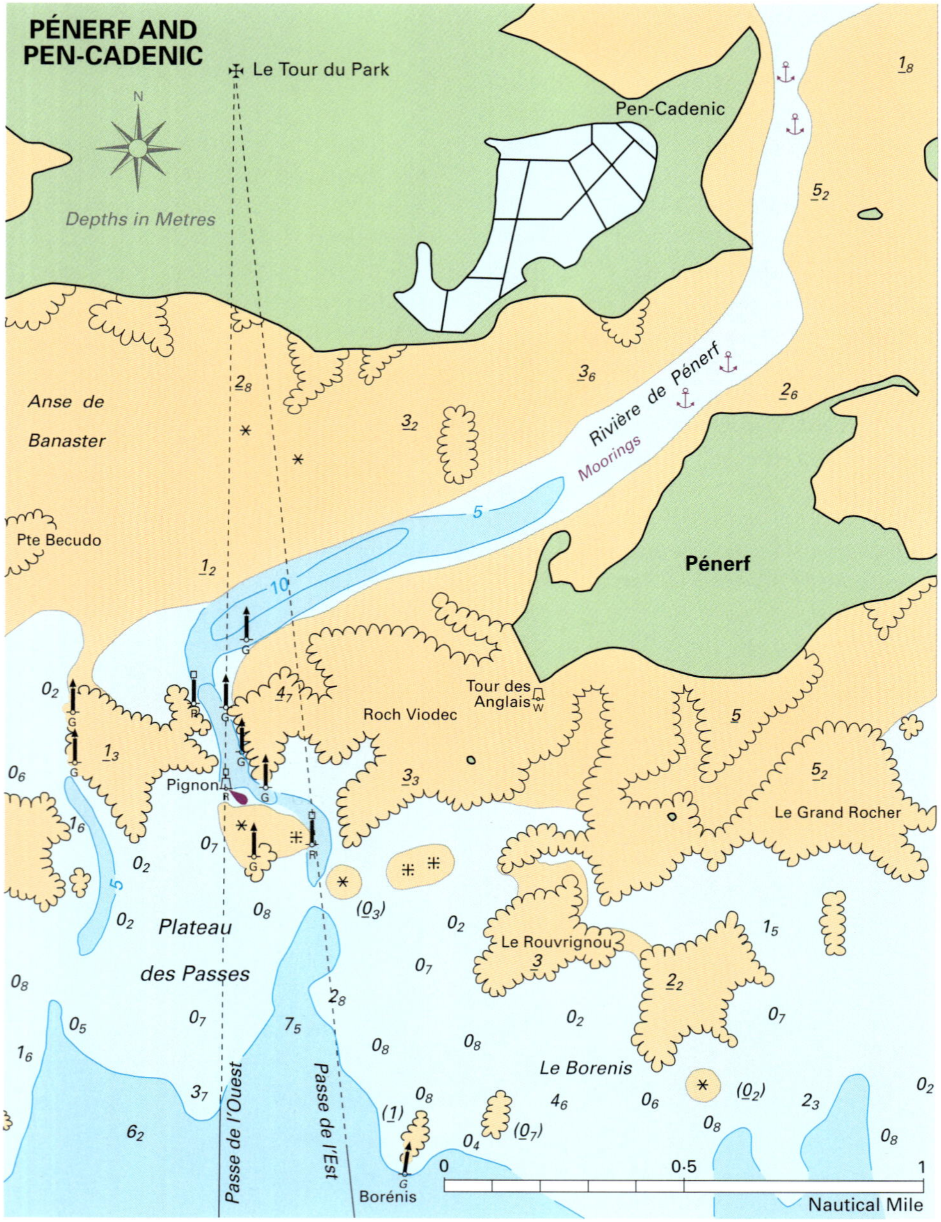

HOUAT AND HOËDIC TO THE VILAINE ESTUARY

KERVOYAL

Some four miles east of Pénerf entrance, Pointe de Kervoyal forms the west arm of the mouth of the Vilaine River. The shallow Anse de Kervoyal is tucked just behind this headland and offers a snug anchorage in westerlies or northwesterlies. The pilotage is straightforward as you are approaching the Vilaine from the southwest: simply leave Basse de Kervoyal S Cardinal beacon tower about half a mile to port and then follow round to the northwest into the bay. Fetch up just outside the local moorings.

The medium-scale Admiralty chart No. 2823 is perfectly adequate for the Vilaine entrance and for approaching Anse de Kervoyal.

Start the approach two hours before high water from a position about three cables south of Basse de Kervoyal S Cardinal beacon tower. From here head northeast towards Bertrand green beacon tower, just over one and a half miles away to the west of Pointe de Penlan. Leave Bertrand to starboard and then follow round to the east and southeast leaving a curving line of green spar beacons to starboard. The quay is to starboard as you come in, but the river soon turns inland to the NNE. Use French SHOM chart No. 7144.

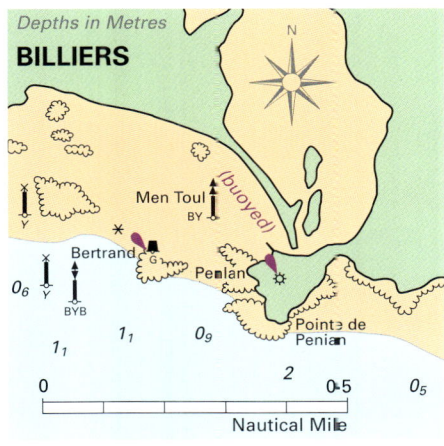

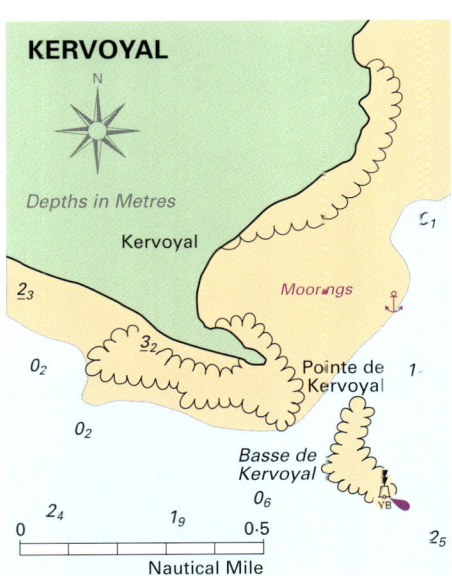

BILLIERS

This narrow tidal harbour lies two miles east of Kervoyal. Although keel boats so inclined can take the ground alongside the quay just inside the entrance, Billiers is included here mainly for the benefit of those with bilge or lifting-keelers with a taste for drying out in unusual havens.

USEFUL ADMIRALTY CHARTS

No. 2823 Quiberon to Croisic
No. 2835 Approaches to Houat and Hoëdic

USEFUL SHOM CHARTS

No. 7135 Pointe de St Jacques à Damgan
No. 7143 Iles de Houat et de Hoëdic
No. 7144 La Vilaine de Damgan à La Roche Bernard

SECRET ANCHORAGES OF BRITTANY **267**

The landing slip and moorings on the south side of Ile d'Arz

CHAPTER 9

THE VILAINE TO THE LOIRE

This final chapter of anchorages looks at the last stretch of Brittany coast between two very different river estuaries – the Vilaine and the Loire. The shallow mouth of the Vilaine can be irascible in a fresh southwesterly, especially on the ebb, but you don't have to venture far upstream before tranquillity prevails and there are tempting places to drop the hook. Only two and a half miles above Tréhiguier, the Arzal barrage effectively canalises the Vilaine and there are plenty of peaceful anchorages thereafter.

270 SECRET ANCHORAGES OF BRITTANY

The winding Vilaine

The 15 miles of coast to the south of the river entrance are rarely cruised by visiting yachts, not really being on the way to anywhere in particular. However, you can find several possible anchorages between Pointe du Halguen and Le Croisic, mostly open to the west but snug in anything from the east.

Round the corner from Le Croisic, the coast is rather exposed to onshore swell, but there are one or two quiet corners near Pointe de Pen Château where you can lie overnight and escape the crowds. A few miles east of the huge marina at Le Pornichet, the Loire estuary marks the end of the Brittany coastline. The lower reaches of the Loire soon turn hostile in fresh onshore winds, belying those placid images of elegant châteaux one associates with this great river. However, I have included four possible anchorages that can be interesting and worth exploring when the weather is right.

The pilotage between the Vilaine and the Loire is not difficult by Brittany standards, and yet each headland that you pass on the way has its own local dangers. Off Pointe du Castelli, the first significant headland south of the Vilaine entrance, the rocks and shoals of Plateau de Piriac spread out for a good one and a half miles to the north and northeast. To the west, the ledges known as Les Bayonelles extend about three-quarters of a mile off Castelli, marked at their limit by a W Cardinal buoy.

The tiny island of Ile Dumet, a recognized bird sanctuary, lies three miles northwest of Pointe du Castelli. I have included a fair weather anchorage

Local barge on the lower Vilaine river

The distinctive Vilaine barrage lock at Arzal

off Dumet's east coast, but there are some dangers to bear in mind if you are just sailing past for a look see. Plateau de l'Ile Dumet is quite a large area of rocky shoals extending north and east from the island for up to one and a half miles. The shallowest heads, about half a mile ENE of the lighthouse, have less than a metre over them at datum. There are also various drying rocks up to four cables east by south of the lighthouse point.

Pointe du Castelli forms the northern arm of the five-mile sweep of Rade de Croisic. The southern headland of this bay, Pointe du Croisic, has some rocky dangers off its north and northwest sides which are cleared, if you are simply rounding the point from the north or south, by passing outside Basse Castouillet W Cardinal buoy.

About three and a half miles west of Pointe du Croisic lurks the potentially dangerous Plateau du Four, marked by a lighthouse at its north end and by a N and a S Cardinal buoy. Le Four is no problem in reasonable visibility, but give it an extra wide berth if conditions are at all murky.

The six straight miles of coast between Pointe du Croisic and Pointe de Penchâteau are mostly steep-to, except for a couple of rocky heads over Basse Lovre about half-way along. These patches, with less than a metre over them, extend half a mile offshore and are marked, rather perversely, by a W Cardinal buoy. La Blanche rocks lie well offshore, five miles to the south of Basse Lovre, but pose no threat if you are coasting close in.

Finally, a long tail of rocks straggles out for three miles southeast of Pointe de Penchâteau, partly sheltering the shallow bay now somewhat dominated by Pornichet marina. Entering this bay from the west, you pass through a narrow gateway of two lateral buoys just off Penchâteau, after which the pilotage is straightforward.

The wide estuary of the Loire, like that of all the grand French rivers, is well littered with drying sandbanks in a kind of delta formation. The main shipping channel – La Passe des Charpentiers – is well buoyed, leading close in towards the west shore of the estuary from the SSW and more or less following this side of the river as far as St Nazaire. Yachts are well advised to follow this channel, perhaps keeping just the wrong side of the buoys if traffic is heavy. The channel has plenty of water at any state of tide, but the streams are strong and you need to carry them in your favour. It's best to enter the estuary on the second half of the flood and leave soon after high water. The Loire should be avoided in fresh southwesterlies, when the approaches are decidedly rough-going.

LA VILAINE

Entering the Vilaine is straightforward in moderate weather above half-flood, but the mouth is shallow and can be rough in strong winds from between west and south, especially on the ebb. Be sure to give a wide berth to La Varlingue, a small rocky shoal (drying 0.3m) lurking half a mile west of Pointe du Halguen, the headland which forms the south arm of the entrance. The north arm is Pointe de Penlan just opposite, a mile to the north. Wide sand-flats extend from either shore for the first couple of miles above Penlan and Halguen, so it's important to follow the buoyed channel. The river narrows between Pointe du Scal and Pointe du Moustoir and then Tréhiguier village appears on the south bank.

Tréhiguier This attractive spot two and a half miles into the river is a useful anchorage, close enough to the mouth to be handy for passage-making and far enough upstream to be fairly sheltered in most conditions. You get an uncomfortable chop in strong westerlies when the tide is running out, but the holding is good in soft mud. Fetch up well clear of the moorings and buoy the anchor. You can land at the slip and there are a few shops and a restaurant ashore. Admiralty chart No. 2823 is adequate for Tréhiguier but the large-scale French SHOM chart No. 7144 is better and well worth having for the Vilaine.

Sailing on the Vilaine

La Grée In westerlies you can find better shelter about a mile further upstream off the south bank, opposite a low stretch of marshy shore near the hamlet of La Grée-Kerdrean. If you tuck in as close as your draught allows, the southward curve of the river gives protection from the west and the holding is good in soft mud. This peaceful spot is more remote than Tréhiguier, but the small town of Camoël is only two kilometres inland.

Les Magues There is an anchorage off the north shore of the Vilaine, not far

SECRET ANCHORAGES OF BRITTANY   273

THE CANALISED VILAINE

Right: The river Vilaine barrage at Arzal

Below: The Pont Cran control tower

Below top: 'Pont Cran' opening bridge, a few miles down from Redon
Below bottom: Vilaine visitor pontoon

The outer approaches to the Vilaine river are wide and shallow, with various rocky dangers on the west side off Penerf. It's best to arrive soon after half-flood when going in, and close to high water when coming out. In fresh onshore winds, especially from the southwest, the mouth of the Vilaine should be avoided. A spring ebb in these conditions kicks up a devilish sea.

About five miles upstream from the entrance you reach the barrage lock at Arzal. Just above the lock are the two large marinas at Arzal-Camoël, but barely a mile beyond this intensive yachting activity you'll start to find plenty of room to anchor in what is effectively a long tideless waterway. When the Vilaine barrage was built, the 42 kilometres of canalised river from Arzal up to Redon yacht basin became easily navigable by yachts with their masts up. With the Atlantic tides held at bay, you can cruise peacefully through rural Brittany with just over four metres minimum depth and 25 metres headroom.

About eight kilometres above Arzal you reach the river port of La Roche Bernard, with its two marinas on the left bank. Some rather striking cliffs rise sharply behind La Roche Bernard and a main road crosses the gorge on a high suspension bridge. The Vieux Port is the traditional demasting or remasting point for a passage through the Brittany canals to Dinan and St Malo, although these waterways are nowadays very shallow and can only be used by bilge-keel yachts or small motor cruisers drawing less than about 1.2 metres. The charming old town of La Roche Bernard has a fascinating medieval quarter, where you can drift back in time wandering narrow streets and secret alleys. In the Château des Basses-Fosses you'll find the Vilaine Maritime Museum with its evocative reconstruction of life in the river and estuary at the turn of the 19th-20th centuries.

Upstream from La Roche Bernard the Vilaine is attractively wooded above rocky shores as far as Foleux village, where the river turns east and the country becomes flatter with more open pastures on either side. You pass several slipways where ferries must once have plied their trade. Between La Roche Bernard and Foleux there are some good spots for anchoring, especially in the shallow mouths of various drowned river creeks that once dried out on the tide but which now have water all the time. Some 13 kilometres above Foleux you reach the swing-bridge at Cran, where there are some local moorings and a waterside restaurant. Just round the next bend is Rieux, with more lines of moorings before you enter the old river port at Redon.

The yacht basin at Redon lies in a sheltered spur off the main river that leads on through a lock to the Canal de Nantes à Brest. Shallow-draught boats with their masts down can follow the Vilaine for another 90 kilometres to Rennes. However, the clearance along this stretch is only 3.2 metres at normal water levels, which can be much reduced if there's an unusually high tide at sea or if plenty of river water is running downstream after heavy rains inland. From Rennes the Canal d'Ille-et-Rance meanders north towards Dinan and the River Rance, although this picturesque Breton waterway has even less clearance, with 2.75 metres in the centre of the older arched bridges and barely 2.3 metres at their sides. The Vilaine at Redon is lined with 17th and 18th-century houses along Quai Duguay-Trouin, stately residences once belonging to wealthy merchants and shipowners.

Around the Vieux Port you can also see several old shipbuilders' houses whose ground floors were once used as offices. The narrow streets around Redon's abbey church of St Sauveur are crowded with well preserved half-timbered buildings leaning together, some dating back to the 15th century. Not far from St Sauveur in Le Cours Clemenceau, keen *boules* players gather in the late afternoons, when the murmur of French voices and the click of the *boules* affirm that all is well with the world.

Among gourmets, Redon is famed as the capital of the sweet chestnut or *marron*. At the end of October each year, a great chestnut fair, the spectacular Foire aux Marrons, is held in the town and you can buy chestnuts in all their creative manifestations – grilled, blended into savoury terrines or made into succulent stuffing for roast pork.

Above: Redon locks

Above left: The suspension bridge at La Roche Bernard

The old barge quays up at Redon

Views of the glorious Vilaine river up at La Roche Bernard

below the Arzal barrage at the mouth of a shallow inlet in the crook of the river bend. This is a useful spot if you have come upstream in the late evening and missed the last lock through the barrage.

La Vieille Roche This once-secluded anchorage, just below the barrage off the south bank, is now packed with moorings. There is still some room to fetch up outside the trots, or you can sometimes find a vacant buoy. La Vieille Roche is perfectly sheltered in strong southwesterlies and very handy for the lock.

Above Arzal There is no room to anchor for the first mile above the Arzal barrage, where both sides of the river are almost completely given over to marina berths, boatyards and moorings. Thereafter, there are plenty of possible anchorages in what is effectively a tideless lake – you simply choose a likely spot and edge towards the bank with an eye on the echo-sounder.

The pontoon berths at La Roche Bernard are very sheltered, whether you opt for the Port Neuf, on the river bank just below the first road bridge, or the rather sleepy Vieux Port in the inlet a little further downstream. However, this attractive stretch of the Vilaine river has become much noisier since the second road bridge was built just upstream from the first to carry the heavy traffic of the N165 dual carriageway.

There are various anchorages in the river beyond La Roche Bernard, of which Foleux is one of the most picturesque, just over three miles above the upstream bridge off the north bank.

LE PILAI

From Pointe du Halguen, the south headland of the mouth of the Vilaine River, the coast trends southward in a series of shallow bays. About one and a half miles south of Pointe du Halguen is a small bay called Le Pilai, just north of Pointe de Marescle and Ile de Belair. There is an anchorage here in quiet or easterly weather, with Ile de Belair bearing SSW about half a mile distant, or a little closer inshore if your draught and the tide allow.

The approach is straightforward, so long as you give a wide berth to Pointe de Loscolo and its off-lying rocky shoals if coming from the south. Don't cut inside Ile de Belair, because there are drying rocks in the whole area between Pointe de Marescle, Pointe de Loscolo and the island. Although it's not safe to enter the anchorage at night, you can leave by heading WNW until you pick up the sectors of Penlan light. If the wind should shift onshore and you are forced to clear out, the Vilaine is not far away. Admiralty chart No. 2823 is adequate for Le Pilai, but the large-scale French SHOM chart No. 7144 is preferable.

Le Pilai

THE VILAINE TO THE LOIRE

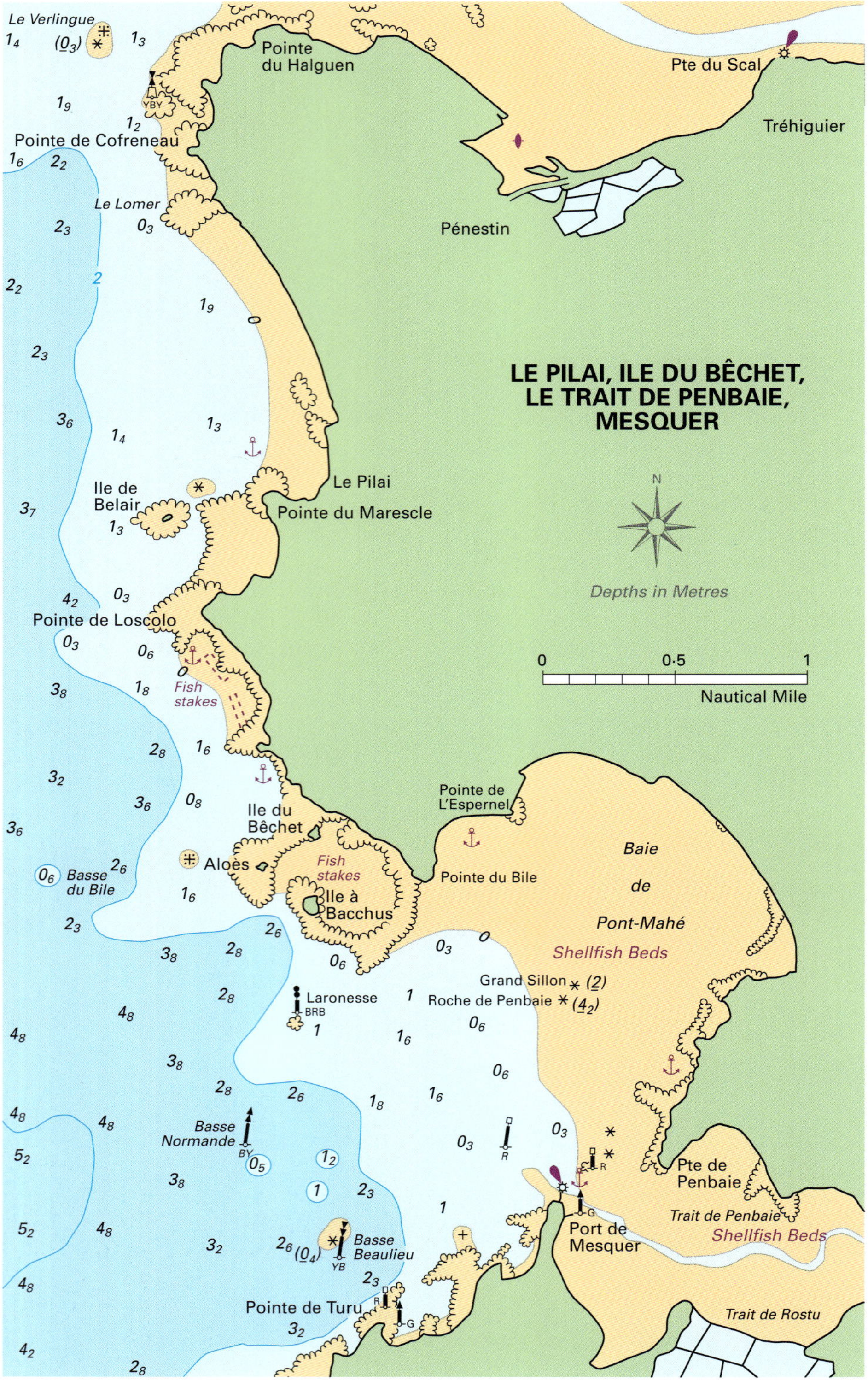

Le Pilai
Ile du Bêchet
Le Trait de Penbaie
Mesquer

LE PILAI, ILE DU BÊCHET,
LE TRAIT DE PENBAIE,
MESQUER

Depths in Metres

THE VILAINE TO THE LOIRE

Loscolo anchorage

ILE DU BÊCHET

Just over a mile south along the coast from Ile de Belair there are three more small islands – Ile du Bêchet, Ile Aloès and Ile à Bacchus, of which Ile du Bêchet is the most northerly and the closest to the shore. All are fringed with drying rocks and there are extensive rocky ledges between the islands and the mainland. In quiet or easterly weather, and provided there's no swell, you can anchor off a small beach a quarter of a mile northwest of Ile du Bêchet.

This is a little used anchorage, well away from the madding crowd. Having identified the three islands, approach the beach from a little north of west, making sure you avoid the isolated rock, awash at datum, which lurks two cables west of Ile Aloès. Near low water watch out for Basse du Bile, with a least depth of 0.6m, about three-quarters of a mile west of Aloès.

You can leave the anchorage at night by heading WNW until you pick up the sectors of Penlan light. The mouth of the Vilaine is less than four miles away. Admiralty chart No. 2823 is adequate for this anchorage, but the French SHOM chart No. 7144 is better.

Ile du Bêchet

SECRET ANCHORAGES OF BRITTANY   279

Ile de Belair

BAIE DE PONT-MAHE (OR LE TRAIT DE PENBAIE)

This shallow marshy inlet is about a mile square and lies just round the corner from Ile du Bêchet, between Pointe du Bile and Pointe de Penbaie (sometimes spelt Pointe de Pen-Bé). Although Baie de Pont-Mahé mostly dries at LAT, there are two neap-tide anchorages where boats of modest draught can just stay afloat, one sheltered from northerlies and the other from easterlies. Because an extensive area of mussel beds covers the entrance to the inlet, you need to approach at high water and reach one of the anchorages before the tide starts falling. Coming into the bay, keep well south of the extensive rocky ledges off Pointe du Bile. Also avoid an isolated drying rock (marked by Laronesse BRB spar beacon) three cables southwest of these ledges, and two drying rocks near the middle of the entrance: Grand Sillon (marked by a spar beacon) and Roche de Penbaie (dries 4.2m).

Approach from due west between Basse Normande N Cardinal buoy to starboard and Laronesse beacon to port. Once you are past Laronesse, head ENE into the bay. When Pointe du Bile bears due north true, you have a choice of turning left or right.

For shelter from the north, come to port towards Pointe de l'Espernel,

Pointe du Bile

Ile à Bacchus

keeping Le Leste church spire fine on your port bow but finally turning NNW to fetch up in the bay between Pointe du Bile and Pointe de l'Espernel. For shelter from the east or southeast, make for the small cove that lies three and a half cables north of the tip of the low promontory, Pointe de Penbaie.

Baie de Pont-Mahé is an interesting natural haven to visit provided the weather is fair and settled, but you need to clear out if the wind should shift to the west or southwest. Although it's not safe for strangers to this inlet to enter at night, you can escape with care if necessary (at high water) by using careful clearing bearings on Pointe de Mesquer light. Refer to Admiralty chart No. 2823 or, preferably, the clear, large-scale French SHOM chart No. 7136.

THE VILAINE TO THE LOIRE

PORT DE MESQUER

This small drying harbour lies close east of Pointe de Mesquer and less than half a mile southwest of Pointe de Penbaie. The local boats based here are partly protected from the west by a jetty extending north from Pointe de Mesquer. There's an attractive anchorage near this breakwater at neaps, in a secluded part of Brittany that few yachts visit.

On the approach, leave Basse Normande N Cardinal buoy close to the south and then steer ESE towards the jetty head, leaving a small red buoy close to port. Round the jetty a cable off, entering the harbour area between it and a red spar beacon. At neaps, in quiet or easterly weather, yachts with a moderate draught can anchor east of the jetty head, outside the local moorings.

Mesquer is a restful spot, nicely off the beaten track, and there are a few shops in the village two kilometres inland. The anchorage is fairly snug in a southwesterly, but exposed to winds from between WSW through west to north. Strangers should not approach in fresh onshore weather or at night, but you can leave at night if necessary by using one of the white sectors of Mesquer light. Note that Basse Normande buoy is unlit. Refer to Admiralty chart No. 2823 or the French SHOM chart No. 7136.

PIRIAC-SUR-MER

Once a small drying harbour, Piriac has grown substantially to encompass a 500-berth marina, accessible through a sill-gate for about two hours each side of high water. The harbour entrance lies four miles southwest along the coast from Pointe de Mesquer and not quite a mile northeast of Pointe du Castelli. The drying outer harbour, just west of the marina, is full of local boats. Piriac was included as an anchorage in the first edition of this book, but few cruising yachts will probably anchor in the offing here now except while waiting for sufficient tide to enter the marina. The sketch chart shows the lines of approach from seaward and from the direction of the Vilaine.

Port de Mesquer

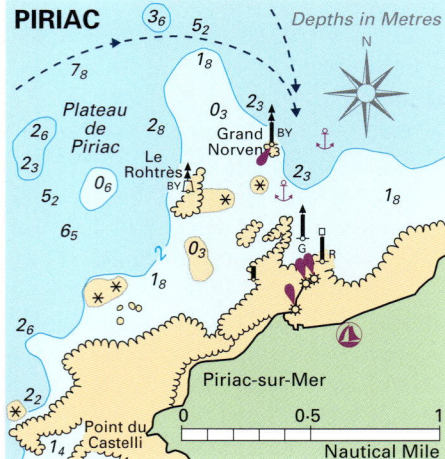

282 ⚓ Secret Anchorages of Brittany

THE VILAINE TO THE LOIRE

ILE DUMET

This low, rather barren island is only a mile long and lies seven miles southwest of the mouth of the Vilaine. Its sole inhabitant is the warden who protects the interests of the bird sanctuary established there. Ile Dumet is fringed by various drying rocks and by the shallow Plateau de l'Ile Dumet, but you can approach from the northeast and anchor off the east shore in quiet settled weather.

Before you are two miles off, bring Ile Dumet lighthouse (near the east tip of the island) to bear 215° and approach on this line. Don't let the lighthouse bear more than 215° or you will come too close to the shallowest parts of the plateau. The shore on this side of the island is fairly clean and you can fetch up about a cable off, with the lighthouse bearing southwest. This anchorage is easy to leave at night. Use Admiralty chart No. 2823 or the French SHOM chart No. 7136.

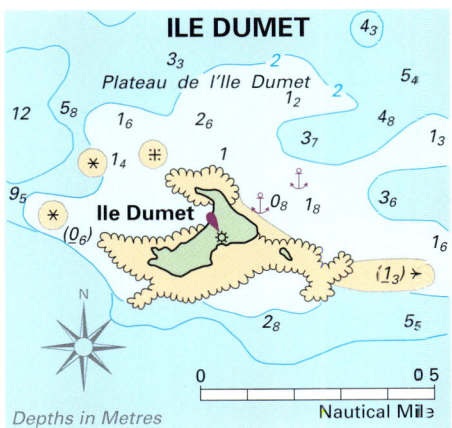

RADE DE CROISIC

This broad bight is three miles across and faces west between Pointe du Castelli and Pointe du Croisic. At the head of the Rade, between the two harbours of La Turballe and Le Croisic, there is a long beach clear of dangers and towards which the depth shelves gradually. You can anchor off this beach in any winds from the east, provided there is no onshore swell.

The approach is straightforward from more or less due west, so long as you avoid the various rocky shoals off Pointe du Castelli or Pointe du Croisic by passing outside Les Bayonelles or Basse Castouillet W Cardinal buoys respectively. One convenient spot to fetch up is about a quarter of a mile southeast of the entrance to La Turballe harbour, off the Plage des Brebis. You can easily reach or leave this anchorage at night, using the white sector of the main breakwater light, but be sure to give a wide berth to the drying rocks that extend south from the east breakwater for well over 100 metres. Refer to Admiralty chart No. 2823 or the French SHOM chart No. 7145.

LA GOVELLE

Rounding Pointe du Croisic for the Chenal du Nord, you come across a six-mile stretch of low indented cliffs and small rocky coves. At its east end this attractive coastline terminates in Pointe de Penchâteau, around and to the north of which lies the entrance to Le Pouliguen. But just over one and a half miles WNW of Penchâteau is a narrow

La Govelle

THE VILAINE TO THE LOIRE

RADE DE CROISIC

inlet known as La Govelle, where you can anchor in winds from between north and northeast, provided there's no onshore swell rolling in.

The final approach is from the WSW, making a slight angle to the coast to avoid two isolated rocks off Pointe de Vicherie, which forms the east side of the inlet. Coming from Pointe du Croisic, pass south of Basse Lovre W Cardinal buoy, from which La Govelle lies one and a half miles to the east. Coming from Penchâteau, give Pointe de Vicherie an offing of three cables before turning into La Govelle. The anchorage is safe to leave at night by heading WSW until you are clear of the coast and Basse Lovre. Refer to French SHOM chart No. 7145.

THE VILAINE TO THE LOIRE

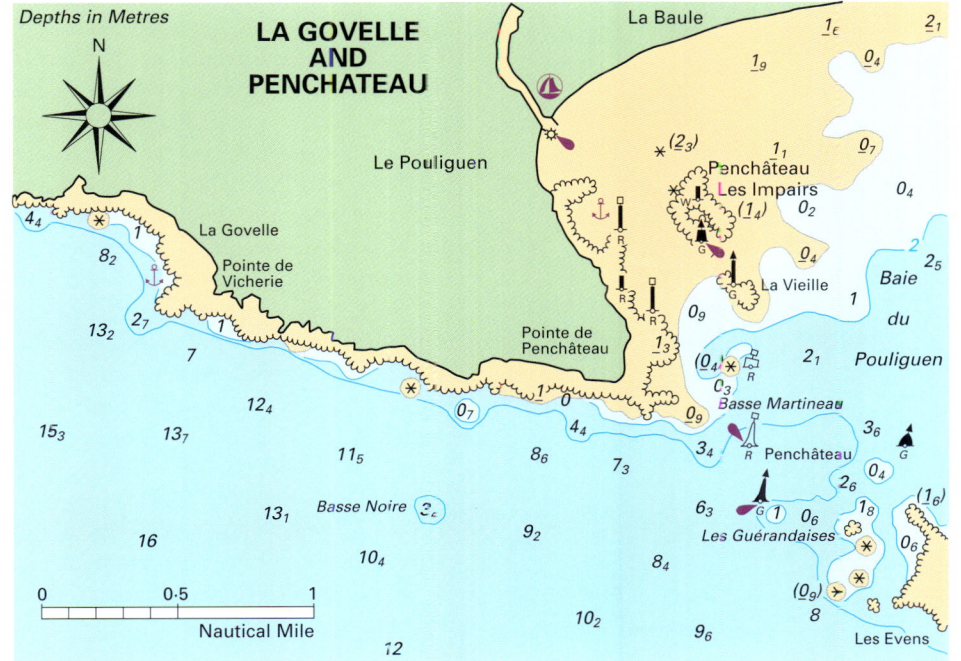

Rade du Croisic

SECRET ANCHORAGES OF BRITTANY 285

THE VILAINE TO THE LOIRE

Small drying harbour on the Côte Sauvage

Penchâteau

PENCHÂTEAU

If you can take the ground easily, there is an attractive drying anchorage half a mile north of Pointe de Penchâteau, just west of the narrow entrance channel that leads into Le Pouliguen. This spot is preferable to the crowded harbour during high season and is well protected from winds with any west in them. You can pick your way in easily enough using Admiralty chart No. 2823, but the pilot books cover Le Pouliguen harbour, which is tucked into the northwest corner of a wide bay cordoned by a string of drying reefs. You need to enter near high water.

The simplest approach is from the west, passing between Penchâteau red buoy and Les Guérandaises green. Then head north to leave Basse Martineau red

286 ⚓ SECRET ANCHORAGES OF BRITTANY

Penchâteau at HW

buoy to port before turning northwest to leave La Vieille green spar beacon and Petits Impairs green beacon tower to starboard, and two red spar beacons to port. The anchorage lies to port after the second red beacon. Entering or leaving at night is not advisable, but Le Pouliguen is close if the wind should shift. The most useful chart is the clear, large-scale French SHOM chart No. 7145.

PORNICHET

The huge yacht harbour at Pornichet is a veritable nautical parking lot, with every conceivable facility and yachts packed cheek by jowl. In quiet weather near neaps, if you fancy a little more seclusion, you can anchor in the bay

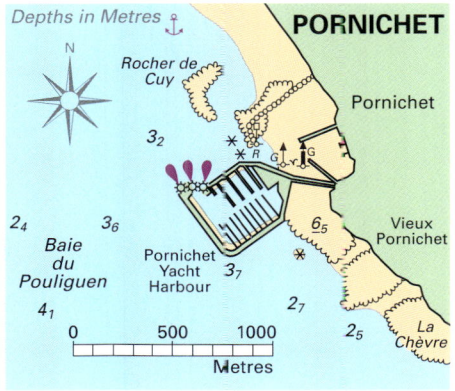

about half a mile north by west of the marina entrance. Approach from Penchâteau buoy as though you were bound for the marina, but then steer for the beach so as to pass well north of the outer breakwater head. Keep an eye on the echo-sounder as you come in, because the bay is very shallow close inshore. Fetch up opposite the large hotel towards the northwest end of Pornichet sea front.

The off-lying reefs keep out much of the swell that would otherwise be sent in by a sea breeze, although you will usually find a persistent roll in this gradually shelving bay unless the weather is definitely offshore. If you anchor overnight, it's easy to slip into the marina if the wind should shift. Refer to the French SHOM chart No. 7145.

American WWII memorial statue in the Loire estuary

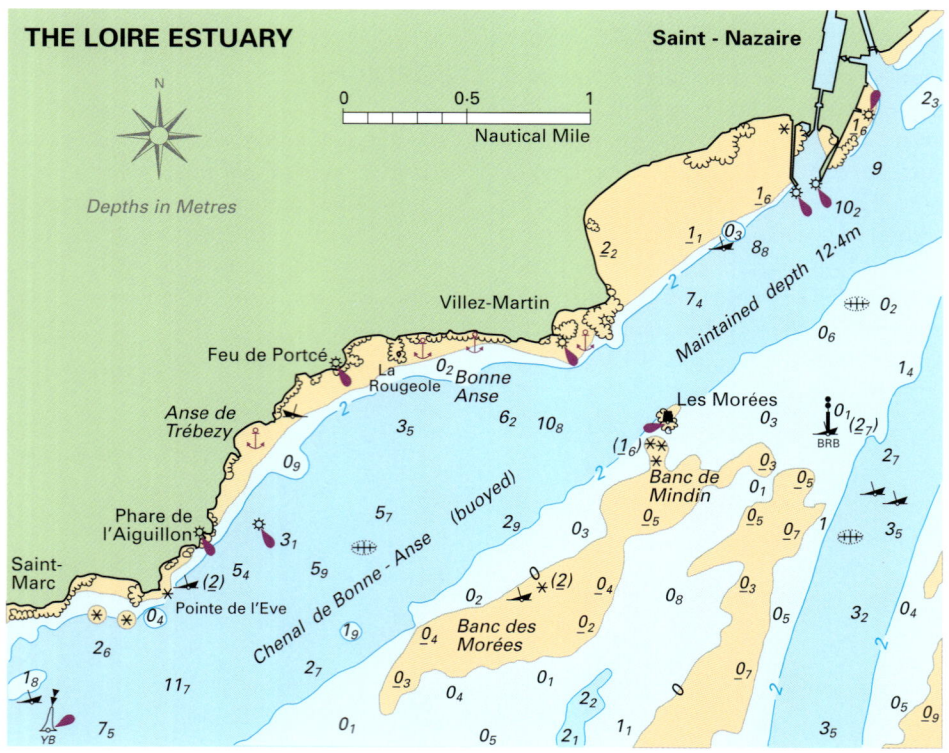

THE VILAINE TO THE LOIRE

THE LOIRE ESTUARY

As you round Pointe de Chémoulin to enter the Loire, the Brittany coast is almost at an end. The Loire is one of the great rivers of France and its estuary, like that of the Seine and the Gironde, can seem rather bleak, even in fine summer weather. The shipping fairway is deep and well buoyed, but there are extensive flats of drying sand across most of the mouth. The main entrance channel hugs the west shore up to St Nazaire, and there are one or two possible anchorages in these lower reaches.

Anse de Trébezy This small bay lies on the west side of the estuary, just opposite Trébezy red buoy and not quite half a mile above Aiguillon lighthouse. The simplest approach is to edge close inshore opposite Aiguillon and then follow the line of the bank to the anchorage. Watch out for Roche Trébezy (awash at LAT) which lies close north of Trébezy buoy, and don't cut too close to the point on the south side of Anse de Trébezy.

There is a spar beacon just outside the drying line at the mouth of the bay,

Anse de Trébezy

which is useful when you are deciding where to anchor. Tuck in as close as you can to avoid the worst of the tide, and set a riding light if you are staying overnight. The anchorage is snug in westerlies or north-westerlies and is easy to leave at night, either to make for the open sea or to venture further up the Loire. The best reference is Admiralty chart No. 2989.

Bonne Anse You'll find a useful anchorage three-quarters of a mile upstream from Trébezy, in the bight between La Rougeole islet and Villez-Martin jetty. A good spot out of the tide

Villez-Martin anchorage

The Loire estuary

SECRET ANCHORAGES OF BRITTANY 289

Traditional Loire fishing hut

The mighty Loire at Paimbœuf, six miles above the St Nazaire bridge

is about one and a half cables east by north of La Rougeole, but avoid the drying rock that lies within a cable northeast of the islet. Bonne Anse is protected from between north and west and is easy to enter or leave at night in reasonable visibility. Refer to the large-scale Admiralty chart No. 2989.

Villez-Martin A small harbour on the west shore of the estuary, opposite Les Morées light-tower and just over a mile below St Nazaire. You can anchor about one and a half cables east by south of Villez-Martin pierhead, clear of the local moorings. This spot is rather more exposed than the previous two anchorages and only offers any real shelter in northwesterlies. Be careful to avoid the drying ledges that extend well out from the pierhead and from the point just north of the anchorage. It's important to set a riding light if you are staying overnight. Villez-Martin is easy to leave at night and is close to St Nazaire. Refer to Admiralty chart No. 2989.

USEFUL ADMIRALTY CHARTS

No. 2823 Quiberon to Croisic
No. 2986 Approaches to La Loire
No. 2989 Entrance to La Loire

USEFUL SHOM CHARTS

No. 6797 Embouchure de la Loire
No. 7136 Baie de Pont-Mahé à Piriac-sur-Mer
No. 7144 La Vilaine de Damgan à La Roche Bernard
No. 7145 La Turballe à Pornichet
No. 7396 Cours de la Loire

INDEX

Aber, Ile de l', 154
Alderney Race, 22, 24, 26
alignements, 152
Aloès, Ile, 279
anchors and ground tackle, 18-19
Anse de Badène, 8, 231, 248, 249-50
Anse de Bénodet, 160, 163, 182-3, 187-8
Anse de Berthaume, 136, 137
Anse de Bréhec, 47, 48
Anse du Cabestan, 171
Anse de Camaret, 19, 126, 149-50, 151
Anse de Combrit, 183, 184-6
Anse de Cornault, 252-3
Anse de Dinan, 151
Anse de Dinard, 37, 39
Anse de Feunteunod, 170
Anse du Fret, 142, 144
Anse de Goërem, 200, 218, 220
Anse de Kerners, 250
Anse de Kernic, 98
Anse de Kersos, 165, 189, 193-4
Anse de Kervoyal, 258, 267
L'Anse de Launay, 59-60
Anse de Lesconil, 162-3, 174-8
Anse du Loc'h, 171
Anse de Locquirec, 86-8
Anse de Morville, 81
Anse de Paimpol, 13, 16, 19, 47-51
Anse du Paux (Pô), 242
Anse de Pen-Hir, 150-51
Anse de Penhap, 250, 251
Anse de Perros, 75-7
Anse Pivette, 24, 26, 27
Anse du Pô (Paux), 242
Anse de Pouldu (Camaret), 150
Anse de Pouldu (Quimperlé), 206
Anse de Poulmic, 144
Anse de St Martin, 24, 26-7
Anse de St Nicolas (Cap de la Chèvre), 153

Anse St Nicolas (Élorn), 142
Anse de St Philibert, 244
Anse de Stervilin, 211
Anse de Stole, 200, 211-13
Anse de Suscinio, 265
Anse de Térénez, 58-9, 88
Anse de Touven, 186-7
Anse de Trébezy, 289
L'Anse de Trestaou, 76, 79-80
Anse de Trestel, 73-4
Anse du Trez, 182-3
Anse de Vauville, 24, 26, 27
Ar Men light, 162, 164
Argenton, 117-18
Argol, 262
artists, 205
Arz, Ile d', 246, 251, 252
Arzal Barrage, 270, 273, 274, 276
L'Auberlac'h, Anse de, 144-5
Audierne, 162, 171, 174
Audierne, Bay of, 168
Aulne River, 122, 124, 144-9
Auray, 248
Auray River, 8, 231, 246-9
Aven River, 199, 201-2, 204, 205

Bacchus, Ile à, 279
Badène, Anse de, 8, 231, 248, 249-50
Baie de Daoulas, 144, 145
Baie de la Forêt, 163-5, 188-94
Baie de la Fresnaie, 13, 43, 44
Baie du Kérogan, 184, 187
Baie de Lampaul, 123-4, 128-9, 132
Baie de Locmalo, 200, 218-20
Baie de Mont-St-Michel, 26, 33
Baie de Pen ar Roc'h, 133
Baie de Pont-Mahe, 278, 280-81
Baie de Pouldohan, 193, 194

Baie de St Brieuc, 26, 62
Baie du Stiff, 123, 129, 132
Baie des Trépassés, 156, 157, 173
Bananec, Ile de, 178, 179
Battle of Quiberon Bay, 234
Batz, Ile de, 55, 96-7, 99-103
Bay of Audierne, 168
Bay of Biscay, 160, 176, 198, 215
Bay of Douarnenez, 123, 126, 151-3, 173
Bec du Colombier, 252
Bêchet, Ile du, 278, 279-80
Beg en Aud, 225
Beg Meil, 188-90, 191
Beg-ar-Vechen lighthouse, 199
Beg-ar-Vir, 213-14
Belair, Ile de, 277, 280
Belle Ile, 216, 228-9, 233-42
Bélon River, 199, 201-2, 204, 205
Bénodet, 10
Bénodet, Anse de, 160, 163, 182-3, 187-8
Bénodet River, 182-3
Berder, Ile, 231, 251
Berthaume, Anse de, 136, 137
Billiers, 258, 267
Binic, 43
birds, 78, 283
Biscay, Bay of, 160, 176, 198, 215
Blavet River, 201, 220, 221
Bodic, 67-9
Bois d'Amour, 205
Bonne Anse, 289
Le Bono, 231, 248, 249
Bono, Ile, 58, 77, 78
Bouches d'Erquy, 43-5
Brebis, Plage des, 283
Bréhat, Ile de, 10, 16, 26, 54, 60-66
Bréhat, Rade de, 63
Bréhec, Anse de, 47, 48
Brest, 141
Brest, Rade de, 122, 124-6

Brigneau, 199, 202-3
Cabestan, Anse du, 171
Callot, Ile de, 17, 87, 90
Camaret, Anse de, 19, 126, 149-50, 151
Camoël, 273
Canal d'Ille-et-Rance, 275
Canal de l'Ile de Batz (Chenal de Batz), 10-11, 96-7, 99-103
Canal de Nantes à Brest, 275
Cancale, 13, 26, 32-3
Cap Carteret, 24
Cap de la Chèvre, 126, 150, 151, 153
Cap Coz, 165, 189, 190
Cap Coz, Plage du, 191
Cap de Flamanville, 24
Cap Fréhel, 13, 43, 44
Cap de la Hague, 12, 22, 26
Carantec, 17, 58, 87, 90, 91
Carnac Plage (Port Endro), 242, 243
Carrec Crom, 108
Carteret, 24
Centre Nautique des Glénans, 178, 180, 251
Cézembre, Ile de, 36-7
Cézon, Ile, 107
Channel Islands, 22-5
charts, 13, 17
Château de Taureau, 88, 89, 90
Châteaulin, 146, 148
Chausey, Iles, 28-32
Chausey Sound, 15-16, 28
Chaussée de Sein, 141, 162, 164
Chenal de Batz (Canal de l'Ile de Batz), 10-11, 96-7, 99-103
Chenal d'Ar-Vas-Du, 165
Chenal d'Ezaudi, 162, 165
Chenal du Four, 99, 123, 138
Chenal Oriental, 162, 166
Chenal des Roquettes à l'Homme, 28-9, 30-31
Chenal de Toulinguet, 150
chestnuts, 275

INDEX

Chevalier, Ile, 182
Chèvre, Cap de la, 126, 150, 151, 153
Cigogne, Ile, 163, 178, 179
Clairfontaine, 26, 27-8
clams, 104
Coastaérès, Ile, 79, 80
cod-fishing, 51
Combrit, Anse de, 183, 184-6
Concarneau, 163-5, 191, 193
Coquilles St Jacques, 62
Cornault, Anse de, 252-3
Corniguel, 184
Corréjou, 106-8, 112
Côte des Abers, 99, 107, 112
Côte de Granit Rose, 54-5
Cotentin coast, 24, 26
Coz, Cap, 165, 189, 190, 191
Coz Pors, 80-81
crabs, 138
Crac'h River, 244
Cran, 274
Créac'h lighthouse, 124, 130
Crevendeilet, 156
Le Croisic, 271, 283
Rade du Croisic, 272, 283, 285
Crouesty, 229, 232, 244, 252, 253
Crozon Peninsula, 151-2

Daoulas, Baie de, 144, 145
Daoulas River, 145, 147
Diélette, 24, 27-8
Dinan, 275
Dinan, Anse de, 151
Dinard, Anse de, 37, 39
Doëlan, 199, 206, 207
Dossen, 97
Douarnenez, 154-5, 156
Douarnenez, Bay of, 123, 126, 151-3, 173
Drénec, Ile de, 178
Drummond Castle, 134
drying out, 17
Dumet, Ile, 271-2, 283

Eckmühl lighthouse, 176
Les Écrehous, 24
Élorn River, 122, 124, 142-3
Erdeven, 220-23
Erquy, Les Bouches d', 43-5
Erquy (St Brieuc Bay), 62
Étel River, 201, 222

Feunteunod, Anse de, 170
fish and fishing
  commercial fishing, 51, 62, 176
  cuisine, 104, 176
  see also *shellfish*
Flamanville, Cap de, 24
Foleux, 274, 276
Le Folgoat, 148, 149
Fôret, Baie de la, 163-5, 189-94
Fort de la Latte, 44
Fréhel, Cap, 13, 43, 44
Frémur River, 41-2
Fresnaie, Baie de la, 13, 43, 44
fruits de mer, 208

Gauguin, Paul, 205
Gavrinis, Ile, 231
Glénan, Iles de, 160, 163, 178-82, 191
goémoniers, 112
Goërem, Anse de, 200, 218, 220
Golchedec, 105
Gorrequer, 145
Goulet de Brest, 141
Grand Harnic islet, 231, 248, 250
Grand Jardin, 36
Grande Ile, 58
Granville, 24
Grève de Feugeot, 252
Groix, Ile de, 200, 212, 213-17
Guéniec, Ile, 107, 111-13
Guéotec, Ile, 178
Guerzido, Port du, 61
Guet, Rade du, 154
Guillec River, 106
Le Guilvinic, 176
Guly Glas, 146, 148
Gwin Zégal, 47

Hague, Cap de la, 12, 22, 26
Hastellic, 261
Havre de Rothéneuf, 17, 35
Les Heaux, 55
Hebihens, Ile des, 42
L'Herbier, 31
Hoëdic, Ile de, 216, 256-7, 258, 262-4
Houat, Ile d', 216, 256, 257, 258, 259-62
Huguenans, Iles des, 30

Ile de l'Aber, 154
Ile Aloès, 279
Ile d'Arz, 246, 251, 252

Ile à Bacchus, 279
Ile de Bananec, 178, 179
Ile de Batz, 55, 96-7, 99-103
Ile du Bêchet, 278, 279-80
Ile de Belair, 277, 280
Ile Berder, 231, 251
Ile Bono, 58, 77, 78
Ile de Bréhat, 10, 16, 26, 54, 60-66
Ile de Callot, 17, 87, 90
Ile de Cézembre, 36-7
Ile Cézon, 107
Ile Chevalier, 182
Ile Cigogne, 163, 178, 179
Ile Coastaérès, 79, 80
Ile de Drénec, 178
Ile Dumet, 271-2, 283
Ile Gavrinis, 231
Ile Grande (Trébeurden), 17
Ile de Groix, 200, 212, 213-17
Ile Guéniec, 107, 111-13
Ile Guéotec, 178
Ile des Hebihens, 42
Ile de Hoëdic, 256-7, 258, 262-4
Ile d'Houat, 216, 256, 257, 258, 259-62
Ile de Keller, 123
Ile du Loc'h, 179, 180
Ile Longue, 231
Ile Losquet, 81-2
Ile Louet, 88, 89
Ile Melon, 16, 118-19
Ile Milliau, 83-5
Ile aux Moines (Morbihan), 250, 251
Ile aux Moines (Sept Iles), 55, 58, 77, 78
Ile Molène (off Trébeurden), 84, 85
Ile de Molène (Ushant), 123, 124, 125, 133-6
Ile aux Moutons, 188, 189
Ile d'Ouessant see Ushant
Ile du Passage (Morbihan), 252
Ile de Penfret, 178, 179-82
Ile de Quéménès, 123, 124, 136, 137
Ile de Raguénès, 195
Ile Renaud, 248
Ile de Rosservo, 107
Ile St Mode, 66-7
Ile de St Nicolas (Iles de Glénan), 179, 181
Ile St Rion, 48

Ile de Sein, 160, 161-2, 164, 165-70
Ile de Siec, 97, 103-6
Ile de Taureau, 90
Ile de Térénez, 146, 149
Ile Teviec, 221, 223, 225
Ile Tomé, 58, 75
Ile Tudy, 182
Ile Valan, 108-9
Ile Vénan, 99, 109-11
Ile Vierge, 99, 106, 108-11, 121
Iles Chausey, 28-32
Iles de Glénan, 160, 163, 178-82, 191
  Centre Nautique, 178, 180, 251
Iles des Huguenans, 30
L'Iroise, 122, 141, 152, 168

Karreg ar Meg, 82-3
Keller, Ile de, 123
Kercanic, 194
Kerlevarac, 248
Kermeur St Yves, 142-3
Kermorvan, Presqu'île de, 128
Kerners, Anse de, 250
Kernic, Anse de, 98
Keroc'h, 200, 210, 211
Kérogan, Baie du, 184, 187
Le Kerpont, 63-4
Kersos, Anse de, 165, 189, 193-4
Kervoyal, Anse de, 258, 267

La Chambre (Bréhat), 10, 60, 61
La Chambre (Iles de Glénan), 178-9
La Chèvre, 64-5
La Corderie, 10, 61, 63, 65-6
La Croix lighthouse, 66-7
La Govelle, 283-4, 285
La Grée, 273
La Horaine, 55
La Jument (Anse de Paimpol), 50
La Jument (Ushant), 123, 129
La Palue, 111
La Passe des Charpentiers, 272
La Roche Bernard, 274, 275, 276
La Roche Jaune, 72
La Roche-Jagu, 69, 71
La Trinité, 229, 243, 244
La Turballe, 283

INDEX

La Vieille Roche, 276
La Vielle lighthouse, 168
La Vilaine, 270, 273-6
L'Aber Benoît, 13, 99, 107, 111, 112, 114-15
L'Aber-Ildut, 99, 112, 119
L'Aberwrac'h, 10, 13, 97, 99, 107, 111, 112
Lagatjar, 152
Lampaul, Baie de, 123-4, 128-9, 132
Landerneau, 143
Landévennec, 145-8
Lannion River, 54-5, 85-6
Lanroz, 183, 184, 187
Lanvaon lighthouse, 108-9
Larmor, 200, 217, 218
L'Auberlac'h, Anse de, 144-5
Launay, L'Anse de, 59-60
Le Bono, 231, 248, 249
Le Croisic, 271, 283
 Rade du Croisic, 272, 283, 285
Le Crouesty, 229, 232, 244, 252, 253
Le Douron River, 86
Le Folgoat, 148, 149
Le Guilvinic, 176
Le Kerpont, 63-4
Le Palais, 233, 239, 240
Le Palus-Plage, 47
Le Passage (Aulne), 144, 146, 149
Le Passage (Élorn), 142
Le Passage (Morbihan), 252
Le Passage (Trieux), 69-71
Le Pilai, 277
Le Pornichet, 271, 286-7
Le Pouldu, 200, 208-11
Le Pouliguen, 286
Le Rocher, 231, 246, 248, 249
Le Ster, 250
Le Stiff, 123, 129, 132
Le Yaudet, 54, 85
Lédénez de Molène, 133-5
Les Heaux, 55
Les Lédénez de Molène, 133-5
Les Magues, 273-6
Lesconil, Anse de, 162-3, 174-8
Lézardrieux, 10, 54, 57, 67
 approaches, 25, 55
 see also Trieux River
L'Herbier, 31
Libenter, 11, 99
L'Ircise, 122, 141, 152, 168

Little Russel Channel, 25
Loc Maria, 200, 214-17
Loc'h, Anse du, 171
Loc'h, Ile du, 179, 180
Locmalo, Baie de, 200, 218-20
Locmariaquer, 230, 244, 259
Locquémeau, 85
Locquirec, Anse de, 86-8
Loctudy, 176, 182
Logonna-Daoulas, 145
Loire Estuary, 272, 288, 289-90
Lomener, 211, 212
Longue, Ile. 231
Lorient, 201, 217, 219, 220
 approaches, 196-7, 199, 200-201, 268-9
Losquet, Ile, 81-2
L'Ost-Pic, 48, 49
Louet, Ile, 88, 89
Les Magues, 273-6

marinas, 8, 9-10
Méaban isles, 244
Melon, Ile, 16, 118-19
Men Brial lighthouse, 162, 165, 166
Merrien, 199, 203-4
Milliau, Ile, 83-5
Minihy-Tréguier, 70
Moguériec, 97, 103-6
Moines, Ile aux (Morbihan), 250, 251
Moines, Ile aux (Sept Iles), 55, 58, 77, 78
Molène, Ile (off Trébeurden), 84, 85
Molène, Ile de (Ushant), 123, 124, 125, 133-6
Mont-St-Michel, 34
Mont-St-Michel, Baie de, 26, 33
Montmarin, Château de, 40
Morbihan, Gulf of, 7, 226-7, 229, 230-32, 244-52
Mordreuc, 39
Morgat, 154
Morlaix River, 10, 13, 55, 58-9, 87-91
Morville, Anse de, 81
Mouillage de Beauchamp, 28-31
Moulin Blanc, Port de, 126
Moutons, Ile aux, 188, 189
mussels, 13, 16, 104, 245
 recipe, 172

Napoleonic Wars, 141
Nez de Jobourg, 26, 27
night safety, 17-18
Nividic lighthouse, 129

Odet River, 183-7
onion johnnies, 93
Orange, Port d', 242
L'Ost-Pic, 48, 49
Ouessant, Ile d' see Ushant
oysters, 13, 33, 104, 208, 231, 245, 248

Paimpol, Anse de, 13, 16, 19
Le Palus-Plage, 47
Passage, Ile du (Morbihan), 252
Paux, Anse du, 242
Pen ar Roc'h, Baie de, 133
Pen Lann, 58, 59, 87, 88-90, 91
Pen Men lighthouse, 213, 214
Pen-Cadenic, 266
Pen-Hir, Anse de, 150-51
Penbaie, Trait de (Baie de Pont-Mahe), 278, 280-81
Penchâteau, 285, 286, 287, 288
Pénerf River, 257, 259, 265, 266
Penfret, Ile de, 178, 179-82
Penhap, Anse de, 250, 251
Penn Enez, 111
Penzé River, 13, 58, 59, 87, 91-2
Perdrix, 69
Perros, Anse de, 75-7
Perros-Guirec, 54, 58, 77, 78
Petit Crom, 108
Petit Minou lighthouse, 141
Le Petit Mont, 244, 251
Le Pilai, 277
pirates, 38, 87
Piriac-sur-Mer, 282
Pivette, Anse, 24, 26, 27
Plage des Brebis, 283
Plage du Cap Coz, 192
Plage du Teven, 187-8
Plouër, 39
Plougonvelin, 136
Plouguerneau, 112
Ploumanac'h, 54, 55, 79, 80
Pô (Paux) Anse du, 242
Pointe de l'Arcouest, 59, 60
Pointe de Beaumer, 242, 243
Pointe du Bec de Vir, 46

Pointe du Bile, 280, 281
Pointe Cameuleut, 13
Pointe du Castelli, 271, 272
Pointe de Combrit, 188
Pointe du Croisic, 272, 283
Pointe de la Croix, 215, 217
Pointe de la Garde, 42, 43
Pointe du Grand Mont, 253
Pointe Halénegui, 263
Pointe du Halguen, 271, 277, 278
Pointe de Kerdonis (Port an Dro), 239
Pointe de la Latte, 43, 44
Pointe de Locquirec, 86-8
Pointe de Loscolo, 277
Pointe de Marescle, 277
Pointe de Pen Château, 271, 272, 283, 285, 286
Pointe de Penmarc'h, 160, 162, 175, 176, 177
Pointe du Raz, 141, 168, 170
Pointe St Colomban, 242-3
Pointe de St Mathieu, 122, 141
Pointe de Toulinguet, 151
Pointe de la Trinité, 59
Pointe du Van, 168
Pointe de la Varde, 35, 36
Pommelin, Rade de, 67
Pont-Aven, 205
Pont-L'Abbé River, 182
Pont-Mahe, Baie de, 278, 280-81
Pontusval, 98, 106
Pornichet, 271, 286-7
Pors Meillou, 184, 186
Pors Moguer, 47
Pors-Poulhan, 174
Port Barrier, 45
Port Blanc (Morbihan), 251
Port Blanc (N Brittany), 16, 54, 73
Port Brenn, 194
Port de la Croix, 257, 262, 263-4
Port an Dro, 239
Port Endro, 242, 243
Port Espagnol, 231
Port la Fôret, 165, 190, 192
Port Goret, 46-7
Port Goulphar, 237-8
Port Guen, 239, 241
Port Halai, 261-2
Port Haliguen, 229
Port Herlin, 238
Port l'Hermite, 202
Port Jean, 241-2

INDEX

Port Kérel, 238
Port Ladron, 250-52
Port Ler, 261-2
Port Manec'h, 199, 201, 205
Port Maria (Aulne), 146
Port Maria (Belle Ile), 239
Port de Mesquer, 282
Port de Moulin Blanc, 126
Port Navalo (Houat), 261
Port Navalo (Morbihan), 231, 244, 245, 256
Port d'Orange, 242
Port du Pouldon, 238-9
Port du Pouldu (Le Pouldu), 200, 208-11
Port de St Gildas, 256, 261
Port St Hubert, 39
Port St Nicolas, 200, 214
Port Salio, 239, 241
Port Tudy, 200, 213
Port du Vieux Château, 233
Port Yorc'h, 239, 241
Portivy, 221, 224, 225
Portrieux, Rade de, 45-6
Portsall, 115-17
Portz Aheac'h, 132
Portz Darland, 132-3
Portz an Dour, 132
Portz Illien, 128
Portz Kernoch, 101-2
Portz Liboudou, 132
Portz Malo, 17, 99, 108-11
Portz Paul, 123, 126-7
Portz Retter, 102-3
Portzmoguer, Anse de, 127, 128
Porz an Ilis, 100-101
Pouldavid River, 155, 156
Pouldohan, Baie de, 193, 194
Pouldon, Port du, 238-9
Le Pouldu, 200, 208-11
Pouldu, Anse de (Camaret), 150
Pouldu, Anse de (Quimperlé), 206
Le Pouliguen, 286
Poulmic, Anse de, 144
Presqu'île de Kermorvan, 128
Presqu'île de Quiberon, 201, 225, 228, 229, 242
Presqu'île Sainte-Marguerite, 107, 111
Quéménès, Ile de, 123, 124, 136, 137
Querelevran Rock, 97, 105

Quiberon Bay, 228-30, 242-3
Battle of, 234
Quiberon Peninsula, 201, 225, 228, 229, 242
Quimper, 184
Quimperlé River, 199-200, 206-11
Rade de Bréhat, 63
Rade de Brest, 122, 124-6
Rade de Cancale, 33
Rade du Croisic, 272, 283, 285
Rade du Guet, 154
Rade de Morlaix, 88
Rade de Pommelin, 67
Rade de Portrieux, 45-6
Raguénès, Ile de, 195
Rance River, 37-40, 275
Barrage, 37, 39, 40
Raz de Sein, 156, 160-62, 168, 173
Redon, 275
Renaud, Ile, 248
Rennes, 275
rocks and reefs, 15, 17
Roscanvel, 144
Roscoff, 93, 103
Rosmeur, 154
Rosservo, Ile de, 107
Rothéneuf, Havre de, 17, 35
Sables d'Or-les-Pins, 43
Sables de Télamot, 190-93
Sables-Blancs (Baie de la Fôret), 190
Sables-Blancs (Douarnenez), 154
St Armel, 252
St Briac-sur-Mer, 39-42
St Brieuc, Baie de, 26, 62
St Cast, 42, 43
St Evette, 174
St Gildas, Port de, 256, 261
St Goustan, 231, 232, 246, 248
St Guénolé, 176
St Hubert, Port, 39
St Jacques-en-Sarzeau, 264-5
St Malo, 10, 36, 38, 41, 42
St Malo, Gulf of, 22-51, 55
St Marguerite, Presqu'Ile, 107, 111
St Marine, 185
St Martin, Anse de, 24, 26-7
St Mode, Ile de, 66-7
St Nazaire, 272

St Nicolas, Anse de (Cap de la Chèvre), 153
St Nicolas, Anse (Élorn), 142
St Nicolas, Ile de (Glénan), 179, 181
St Nicolas, Port (Groix), 200, 214
St Norgard, Anse de, 153, 154
St Philibert, Anse de, 244
St Philibert River, 16, 243
St Quay-Portrieux, 45-6
St Rion, Ile, 48
St Servan, 37
St Suliac, 39
St Yves (Penzé), 92
St Yves of Tréguier, 70
scallops, 62, 104
seaweed gathering, 112
Sein, Ile de, 160, 161-2, 164, 165-70
Sein, Raz de, 156, 160-62, 168, 173
semaphore signalling, 222
Le Sénéquet, 24
Sept Iles, 55, 58, 76, 77, 78
shellfish 13, 104, 138, 245, see also *mussels; oysters; scallops*
shipwrecks, 134
Siec, Ile de, 97, 104-6
Solidor Bay, 37
standing stones, 152
Le Ster, 250
Ster-Vraz, 235-7
Ster-Wenn (Ster-Ouen), 233, 235-7
Stervilin, Anse de, 211
Stiff, Baie du, 123, 129, 132
Stole, Anse de, 200, 211-13
Styvel, 145-9
submarines, 219
Suscinio, Anse de, 265
Swinge, 23
Tas de Pois, 150, 151
Taureau, Ile de, 90
Télamot, Sables de, 190-93
Térénez, Anse de, 58-9, 88
Térénez, Ile de, 146, 149
Teven, Plage du, 187-8
Tévennec islet, 161, 162, 168
Teviec, Ile de, 221, 223-4
tides, 15-17
Tinduff, 145
Tomé, Ile, 58, 75
Toulinguet, Chenal de, 150

Toulinguet, Pointe de, 151
Toulven, Anse de, 186-7
Trait de Penbaie (Baie de Pont-Mahe), 278, 280-81
Tréac'h er Béniguet, 260
Tréac'h er Goured, 254-5, 256, 257, 261
Tréac'h Salus, 260-61
Trébeurden, 17, 54, 58, 85
Trébezy, Anse de, 289
Trébour, 154
Trégarvan, 144, 146, 149
Trégastel, 54, 80-81
Tréguier, 70, 71
Tréguier River, 6, 9-10, 13, 54, 57, 71-2
Tréhiguier, 270, 273
Trépassés, Baie des, 156, 157, 173
Trestaou, L'Anse de, 76, 79-80
Trestel, Anse de, 73-4
Trez, Anse du, 182-3
Trieux River 13, 66-71, see also Lézardrieux
Tudy, Ile, 182

U-boats, 219
Ushant (Ile de Ouessant), 123-4, 128-33, 134, 140, 159

Valan, Ile, 108-9
Vannes, 232, 256
Vauville, Anse de, 24, 26, 27
Vénan, Ile, 99, 109-11
vent solaire, 18, 198-9, 216
Vierge, Ile, 99, 106, 108-11, 121
Vieux Château, Port du, 233
Vilaine River, 258, 259, 267, 270, 273-6
Barrage, 270, 273, 274, 276
Villez-Martin, 289

weather, 17-18
winds, 17-18, 198-9, 216
winkles, 104
World War II, 107, 113, 219

Le Yaudet, 54, 85
Ys (lost city), 173